McGRAW-HILL NETWORKING AND TELECOMMUNICATIONS

Crash Course

Shepard	*Telecom Convergence, 2/e*
Shepard	*Telecom Crash Course, 2/e*
Bedell	*Wireless Crash Course, 2/e*
Shepard	*VoIP Crash Course*

McGraw-Hill Communications

Smith/Collins	*3G Wireless Networks*
Bates	*Broadband Telecom Handbook, 2/e*
Collins	*Carrier Grade Voice over IP*
Benner	*Fibre Channel for SANs*
Bates	*GPRS*
Sulkin	*Implementing the IP-PBX*
Russell	*Signaling System #7, 4/e*
Karim/Sarraf	*W-CDMA and cdma2000 for 3G Mobile Networks*
Bates	*Wireless Broadband Handbook*
Rohde/Whitaker	*Communications Receivers, 3/e*
Sayre	*Complete Wireless Design*
OSA	*Fiber Optics Handbook*
Bates	*Optimizing Voice in ATM / IP Mobile Networks*
Roddy	*Satellite Communications, 3/e*
Simon	*Spread Spectrum Communications Handbook*
Snyder	*Wireless Telecommunications Networking with ANSI-41, 2/e*

Reference

Muller	*Desktop Encyclopedia of Telecommunications, 3/e*
Clayton	*McGraw-Hill Illustrated Telecom Dictionary, 3/e*
Pecar	*Telecommunications Factbook, 2/e*
Russell	*Telecommunications Pocket Reference*

WIMAX HANDBOOK

WiMAX Handbook

Building 802.16 Wireless Networks

Frank Ohrtman

McGraw-Hill

New York   Chicago   San Francisco   Lisbon
London   Madrid   Mexico City   Milan   New Delhi
San Juan   Seoul   Singapore   Sydney   Toronto

The **McGraw·Hill** Companies

3 4 5 6 7 8 9 0   DOC/DOC   0 1 9 8 7 6

ISBN 0-07-145401-2

The sponsoring editors for this book were Stephen Chapman and Jane Brownlow and the production supervisor was David Zielonka. It was set in Century Schoolbook by MacAllister Publishing Services, LLC.

Printed and bound by RR Donnelley.

McGraw-Hill books are available at special quantity discounts to use as premiums and sales promotions, or for use in corporate training programs. For more information, please write to the Director of Special Sales, Professional Publishing, McGraw-Hill, Two Penn Plaza, New York, NY 10121-2298. Or contact your local bookstore.

 This book is printed on recycled, acid-free paper containing a minimum of 50 percent recycled de-inked fiber.

This book is dedicated to my son, Konrad Franklin Ohrtman, born June 2004. May all this be ancient technological history by the time he is old enough to read this book.

ABOUT THE AUTHOR

Frank Ohrtman has almost 20 years experience in VoIP and wireless applications. Mr. Ohrtman learned to perform in-depth research and write succinct analyses during his years as a Navy Intelligence Officer (1981–1991) where he specialized in electronic intelligence and electronic warfare. He is a veteran of U.S. Navy actions in Lebanon (awarded Navy Expeditionary Medal), Grenada, Libya (awarded Joint Service Commendation Medal) and the Gulf War (awarded National Defense Service Medal).

His career in VoIP began with selling VoIP gateway switches for Netrix Corporation to long distance bypass carriers. He went on to promote softswitch solutions for Lucent Technologies (Qwest Account Manager) and Vsys (Western Region Sales Manager). Mr. Ohrtman is the author of *Softswitch: Architecture for Voice over IP*, a number one bestseller on USTA Bookstore's bestseller list, *Wi-Fi Handbook: Building 802.11b Wireless Networks,* and *Voice over 802.11*

He holds a master of science in Telecommunications from Colorado University College of Engineering (master's thesis: "Softswitch as Class 4 Replacement—A Disruptive Technology"), a master of arts in International Relations from Boston University and a BA, Political Science from University of Iowa. Mr. Ohrtman lives in Denver, CO where he is the president of WMX Systems, a next generation networks professional consulting and systems integration firm, (http://www.wmxsystems.com frank@wmxsystems.com) 720-839-4063.

CONTENTS AT A GLANCE

CONTENTS

Contents

Contents

ACKNOWLEDGMENTS

Like any book, this project would not have been possible without the generous assitance of a number of dedicated professionals. I would especially like to recognize Herm Braun, professional engineer, "Wireless Emeritus" at Denver University, for providing a sanity check on the technical chapters. Roger Marks of the IEEE 802.16 Working Group for steering my research in the right direction while waiting for the final specification to be published. Also in the Working Group, a big thanks to Dean Chang, Carl Eklund, Kenneth L. Stanwood, and Stanley Wang. To the Intel team, a big thanks to Govindan Nair, Joey Chou, Tomaz Madejski, Krzysztof Perycz, David Putzolu, and Jerry Sydir. I also want to extend a special thanks to the WiMAX Forum for their help in the economics chapter. Tim Stewart of NetUnwired was especially helpful in guiding me to understand the "real products" aimed at "the real market." I would also like to thank Dan Lubar for his technical support along the way as well as Charlie Loverso and Kevin Suitor of Redline Communications.

PREFACE

I wrote my first book, *Softswitch: Architecture for VoIP*, partially as a treatise on how the telecommunications industry could bypass the incumbent telephone company's central office (CO). That still left "the last mile." I then wrote (with Konrad Roeder as coauthor) *Wi-Fi Handbook: Building 802.11b Wireless Networks*, with an eye to IEEE 802.11b (aka Wi-Fi) as a last-mile wireless, unlicensed bypass of the telco's copper wires. To underline that assertion, I went on to author *Voice over 802.11*.

However, 802.11 technologies lacked the throughput, power, and range to be considered "carrier class" replacements for the copper wire last mile. When I started to study WiMAX (IEEE 802.16), I began to see that it was the final piece that would allow a complete bypass of the telco's public switched telephone network (PSTN). I would not rest until I compiled and published this book. Therefore, this book is a very short treatise on how the PSTN can be bypassed in its entirety.

Some state that WiMAX is overhyped. I disagree. It is built on legacy technologies conforming with Data-Over-Cable Service Interface Specification (DOCSIS). It should be noted that even though the specification was approved only 4 months prior to the time of this writing (Fall 2004) and that true 802.16-spec chips will not be available until mid-2005, there have been a number of WiMAX-like or "pre-WiMAX" products on the market (built by vendors participating in the various 802.16 working groups) that perform close to the parameters of the specification. This should be verification enough of the performance of WiMAX.

Introduction

This book describes the Institute of Electrical and Electronic Engineers (IEEE) standard 802.16, more popularly known as Worldwide Interoperability for Micro Wave Access, or WiMAX. The standard, which was years in the making, was finalized in June 2004. This book will attempt to give a brief technical overview of the standard per the specification, followed by a series of discussions of how the technology can deliver the triple play of data, voice, and video.

WiMAX will change telecommunications, as it is known throughout the world today. It eradicates the resource scarcity that has sustained incumbent service providers for the last century. As this technology enables a lower barrier to entry, it will allow true market-based competition in all of the major telecommunication services: voice (mobile and static), video, and data.

Since the inception of the telephone, service providers have staved off competition by relying on the exorbitant capital investment necessary to deploy a telephone network. The cost of deploying copper wires, building switches, and connecting the switches created an insurmountable barrier to entry for other competitors. In most of the world, the high cost of this infrastructure limited telephone service to the wealthy and the fledgling middle class.

WiMAX offers a point-to-point range of 30 miles (50 km) with a throughput of 72 Mbps. It offers a non-line-of-sight (NLOS) range of

Figure 1-1
WiMAX delivers 72 Mbps over 30 miles point-to-point and 4 miles NLOS.

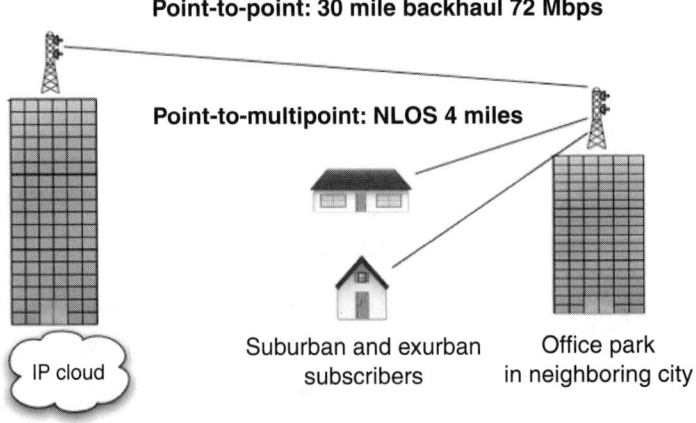

Point-to-point: 30 mile backhaul 72 Mbps

Point-to-multipoint: NLOS 4 miles

Suburban and exurban subscribers

Office park in neighboring city

IP cloud

"Lit" building in metro area

4 miles and, in a point-to-multipoint distribution, the model can distribute nearly any bandwidth to almost any number of subscribers, depending on subscriber density and network architecture. Figure 1-1 illustrates these exciting capabilities.

Telecommunications Networks— The Need for an Alternative Form of Access

An understanding of the workings of the Public Switched Telephone Network (PSTN) is best grasped by understanding its three major components: access, switching, and transport. Each element has evolved over the hundred year plus history of the PSTN. Access pertains to how a user accesses the network, switching refers to how a call is "switched" or routed through the network, and transport describes how a call travels or is "transported" over the network. This network was designed originally to handle voice; later, data was introduced. As data traffic on the PSTN grew, high-capacity users found it inadequate, so these subscribers moved their data traffic to data-specific networks. Many data users then found themselves limited to an infrastructure that was dependent on wires, either fiber optic cable, coaxial cable, or twisted pair copper wire. While wireless means of communication are not new (forms of radio communication have been in use for almost a century), using wireless means to bypass wired monopolies is now a practical opportunity for subscribers of both voice and data services. The primary form of bypass is the use of cellular phones. WiMAX is a wireless technology that holds great promise in delivering broadband (up to 11 Mbps) data.

Figure 1-2
The three components of a telephone network: access, switching, and transport

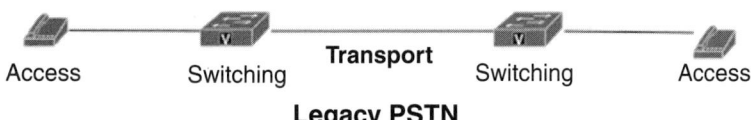

Access   Switching   **Transport**   Switching   Access
Legacy PSTN

Switching

The PSTN is a star network; that is, every subscriber is connected to another via at least one if not many hubs, known as *offices*. In these offices are switches. Very simply, local offices are for local service connection, and tandem offices are for long-distance service. Local offices, better known as central offices (COs), use Class 5 switches while tandem offices use Class 4 switches. A large city might have several COs. Denver (population two million), for example, has approximately 40 COs. COs in a large city often take up most of a city block and are recognizable as large brick buildings with no windows.

Transport

It took more than a century to build the PSTN at great expense. Developers have been obsessed over the years with getting the maximum number of conversations transported at the least possible cost in infrastructure. Imagine an early telephone circuit running from New York to Los Angeles. The copper wire, repeaters, and other mechanisms involved in transporting a conversation this distance were immense. Hence, the early telephone engineers and scientists had to find ways to get the maximum number of conversations transported over this network. Through much research, they developed different means to wring the maximum efficiency from the copper wire infrastructure. Many of those discoveries translated into technologies that worked equally well when fiber optic cable came onto the market. The primary form of transport in the PSTN has been circuit switched (as opposed to the Internet's packet switching). In the 1990s, long-distance service providers, or inter exchange carriers (IXCs), and local service providers, or local exchange carriers (LECs), have migrated those transport networks to asynchronous transfer mode (ATM). ATM is the means for transport from switch to switch. The emergence of Internet Protocol (IP) backbones is drawing much traffic from ATM networks and onto IP networks.

Access

Access refers to how the user accesses the telephone network. Most users gain access to the network via a telephone handset. This handset is usually connected to the CO (where the switch is located) via copper wire known as *twisted pair* because, in most cases, it consists of a twisted pair of copper wires. The stretch of copper wire connects the telephone handset to the CO. One of the chief reasons the majority of subscribers have no choice in local service providers is the prohibitive expense of deploying any alternative to the copper wire that now connects them to the network. Second, gaining right-of-way across properties to reach subscribers would border on the impossible, both in legal and economic terms.

Replacing the PSTN One Component at a Time

The three components of the PSTN are being replaced in the free market via substitution by other technologies and changes in the regulatory atmosphere. The Memorandum of Final Judgement of 1984 (MFJ of 1984) opened the transport aspect of the PSTN to competition. This caused an explosion in the number of long-distance service providers in the United States. The bandwidth glut of 2000 has driven down the cost of long-distance transport.

The Telecommunications Act of 1996 was intended to further the reforms brought on by the MFJ of 1984 but has failed to do so. This act specified how incumbent telephone companies were to open their switches to competitors; however, the incumbents stalled this access first by legal maneuver and second by outright sabotage. They employed the same tactics by blocking competitive access to the access side of their networks. A technology known as *softswitch* offers a technology bypass of the PSTN switches; however, the last mile (aka "the first mile") still remains under the control of the incumbent service providers.

Objections to Wireless Networks

The position that wireless technologies will replace the PSTN meets with a number of objections. Primarily, these objections are focused on quality of service (QoS) issues, security of the wireless network, limitations in the range of the delivery of the service, and the availability of bandwidth. This book will explain how these objections have been overcome.

QoS

One of the primary concerns about wireless data delivery, as with the Internet over wired services, is that the QoS is inadequate. Contention with other wireless services, lost packets, and atmospheric interference are potential objections to WiMAX as an alternative to the PSTN. QoS is also related to the ability of a wireless Internet service provider (WISP) to accommodate voice on its network. WiMAX utilizes a number of measures to ensure good QoS, including service flow QoS scheduling, dynamic service establishment, and a two-phase activation model. Figure 1-3 illustrates broadband wireless as an alternative to the PSTN infrastructure.

Security

WiMAX uses a X.509 encryption to set up the session and, once established, uses 56-bit DES encryption to protect the transmission. Both measures block theft of service and ensure the privacy of the session.

Interference Mitigation

The Radio Act of 1927 has driven the wireless regulatory framework in the United States. It is time for change. The current Federal Communications Commission (FCC) is at least somewhat aware that

Figure 1-3
Overview of a
broadband
wireless
alternative to
the PSTN

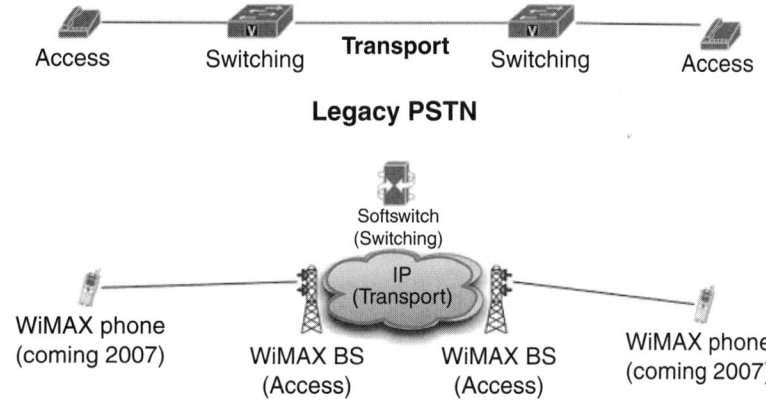

Figure 1-3
Overview of a broadband wireless alternative to the PSTN

wireless poses a third means (after the telephone company's copper wire and the cable TV company's coaxial cable) of delivering residential broadband and that when broadband Internet access is as ubiquitous as land line telephone service is today the U.S. economy can enjoy a $500 billion annual benefit.

Economic Advantage of WiMAX

Wireless technologies potentially pose a cost-effective solution for service providers, in that these technologies do not require right-of-way across private or public property to deliver service to the customer. Many businesses cannot currently receive broadband data services, as no fiber optic cable runs to their building(s). The cost of securing permission to dig a trench through another property and running the requisite cable is prohibitive. With WiMAX and associated technologies, it is possible to merely "beam" the data flow to that building. This solution carries over to the small office/home office (SOHO) market, in that the data flow can be beamed to homes and small businesses in places where no fiber optic or other high-bandwidth service exists.

Regulatory Aspects of Wireless Networks

What are the regulatory concerns when deploying a wireless enterprise network? For a WISP? The FCC addresses wireless services in what is popularly known as Part 15. Wireless data requires a spectrum on which to transmit over the airwaves at a given frequency. An unlicensed spectrum does not require the operator to obtain an exclusive license to transmit on a given frequency in a given region. Unlike the operators of radio stations or cellular telephone companies, a WISP, public or private, is transmitting "for free." Assuming WISPs ultimately compete with cell phone companies for subscribers, WISPs that utilize WiMAX technologies may find themselves at a strong advantage over third-generation networks (3Gs).

Improved Quality of Life with Wireless Networks

When deployed as a broadband IP network solution, WiMAX will enable an improved standard of living in the form of telecommuting, lower real estate prices, and improved family lives. A wave of opportunity for wireless applications is in the making. Most of it lies in the form of broadband deployment. The potential for "better living through telecommunications" lies largely with the ubiquitous availability of broadband. In their April 2001 white paper, *The $500 Billion Opportunity: The Potential Economic Benefit of Widespread Diffusion of Broadband Internet Access,* Robert Crandall and Charles Jackson point to an economic benefit of $500 billion per year for the American economy if broadband Internet access were to be as ubiquitous as land line phones.[1]

---

[1]Robert W. Crandall and Charles L. Jackson. "The $500 Billion Opportunity: The Potential Economic Benefit of Widespread Diffusion of Broadband Internet Access." Washington, DC: Criterion Economics, LLC, 2001. Available at www.criterioneconomics.com.

Disruptive Technology

In his 2000 business book, *The Innovator's Dilemma*, Clayton Christensen describes how disruptive technologies have precipitated the failure of leading products as well as their associated and well-managed firms. Christensen defines criteria to identify disruptive technologies regardless of their market. These technologies can potentially replace mainstream technologies and their associated products and principal vendors. Christensen abstractly defines disruptive technologies as "typically cheaper, simpler, smaller, and, frequently, more convenient" than their mainstream counterparts.[2] WiMAX fits these criteria. Figure 1-4 illustrates this potential disruption as posed to a variety of telecommunications industries. The following industries are threatened with disruption by WiMAX.

Figure 1-4
WiMAX is potentially disruptive to a number of telecommunications industries.

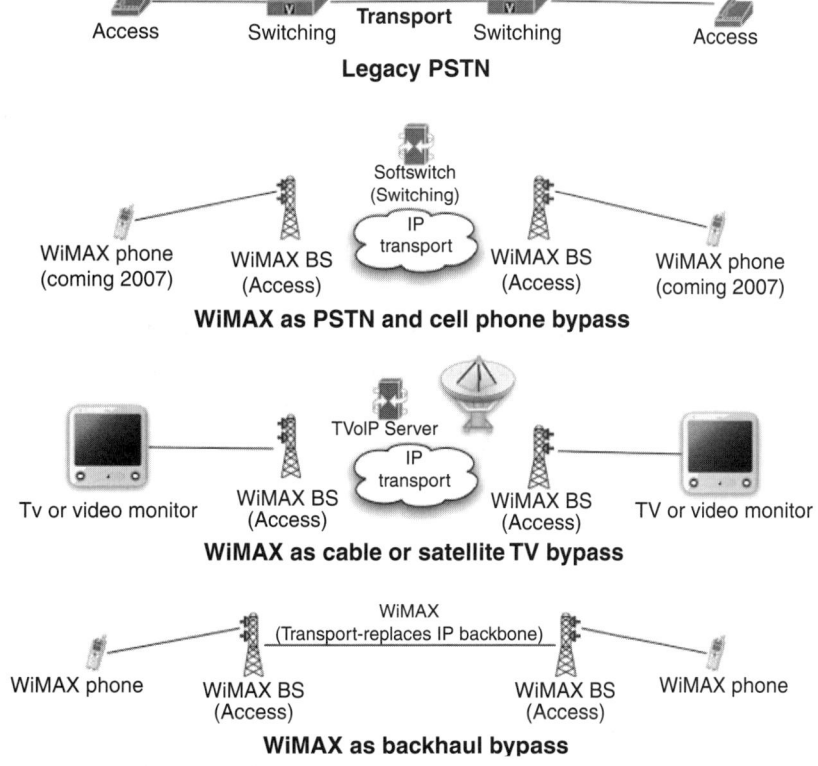

Access   Switching   **Transport**   Switching   Access
Legacy PSTN

Softswitch (Switching)
IP transport

WiMAX phone (coming 2007)   WiMAX BS (Access)   WiMAX BS (Access)   WiMAX phone (coming 2007)
WiMAX as PSTN and cell phone bypass

TVoIP Server
IP transport

Tv or video monitor   WiMAX BS (Access)   WiMAX BS (Access)   TV or video monitor
WiMAX as cable or satellite TV bypass

WiMAX (Transport-replaces IP backbone)

WiMAX phone   WiMAX BS (Access)   WiMAX BS (Access)   WiMAX phone
WiMAX as backhaul bypass

[2]Clayton Christensen, *Innovator's Dilemma,* Harper Business with permission from Harvard Business School Press, New York, NY, 2000, p. 221.

Disruption for Telephone Companies

Figure 1-1 demonstrated how WiMAX replaces the access portion of the PSTN. The broadband Internet connection made possible by WiMAX is IP and, using Voice over Internet Protocol (VoIP), the PSTN is bypassed. With the possible exception of terminating a voice call to a PSTN number, calls need not touch the PSTN. This is potentially very disruptive to incumbent telephone companies. Refer to Figure 1-2 for an illustration.

Disruption for Cable TV and Satellite TV Companies

A technology called TV over Internet Protocol (TvoIP) does for cable TV what VoIP does for telephone companies. It is now possible to simply convert cable TV programming and deliver it over a broadband Internet connection such as WiMAX. The programming is available in real time identical to the cable TV broadcast, and channels can be changed using a set top box while programming is displayed on a conventional TV set. No PC skills are required.

Disruption for Cell Phone Companies

VoIP technologies can be used for mobile telephony to replace incumbent cell phone technologies. It will soon be possible to replace an incumbent cell phone infrastructure for a small fraction of the cost of building the incumbent cell phone network. All that is really necessary is a WiMAX mobile phone and access to a WiMAX base station (the same base stations that deliver broadband Internet access, VoIP, and TvoIP to residences and businesses).

Disruption for the Backhaul Industry

The building of multibillion dollar fiber optic networks marked the telecommunications boom of the 1990s. Very simply put, if WiMAX can beam 72 Mbps over 30 miles and the infrastructure costs only a few thousand dollars (radios, antennas), then services that backhaul (or transport) data via fiber optic cables and charge their customers thousands of dollars per month to do so are in jeopardy. This model can be extended to long-distance backhaul as well. Microwave towers have long been the means of long-distance backhaul for telephone companies. WiMAX is a means of simply expanding or augmenting these networks.

Conclusion

As of 2005, despite the guarantees contained in the Telecommunications Act of 1996, it appears obvious that competition will never come *in* the local loop but can only come *to* the local loop in the form of an alternative network. Consumers will only enjoy the benefits of competition in the local loop when and where alternative technology in switching and access offer a competitor lower barriers to entry and exit in the telecommunications market. If telecommunications consumers are to enjoy the benefits of competition in their local loop, a form of bypass of the switching architecture and the means of access (copper wires from the telephone company) must be offered.

WiMAX: The Physical Layer (PHY)

Introduction

WiMAX is not truly new; rather, it is unique because it was designed from the ground up to deliver maximum throughput to maximum distance while offering 99.999 percent reliability. To achieve this, the designers (IEEE 802.16 Working Group D) relied on proven technologies for the PHY including orthogonal frequency division multiplexing (OFDM), time division duplex (TDD), frequency division duplex (FDD), Quadrature Phase Shift Keying (QPSK), and Quadrature Amplitude Modulation (QAM), to name only a few. This chapter will provide a brief overview of the PHY and different variants (based on their PHY technologies and applications) of WiMAX, the technologies that make these variants work, and reasons why these technologies combine to make WiMAX a quantum leap over previous wireless technologies.

Figure 2-1
IEEE 802.11
MAC and
physical layers
(Source:
McGraw-Hill)

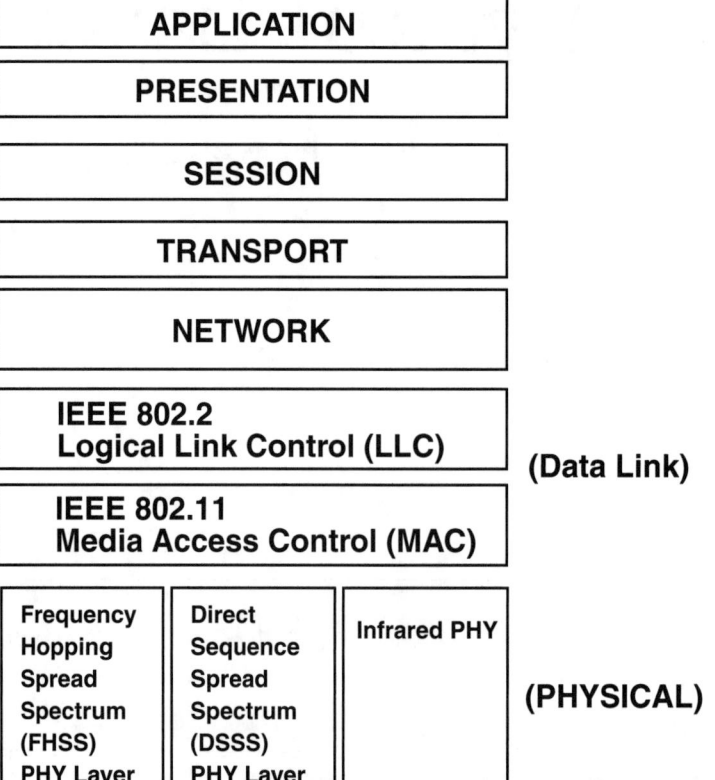

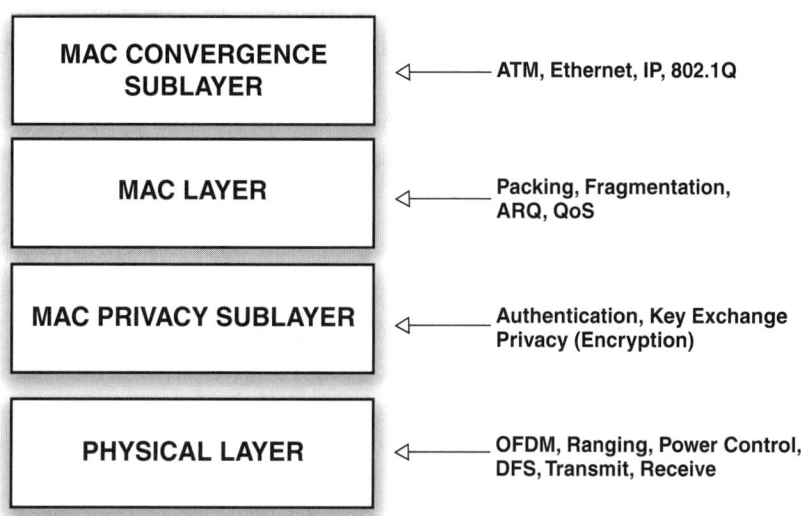

Figure 2-2
MAC and physical layers of IEEE 802.16 as detailed by the IEEE (Source: IEEE)

As the name implies, 802.16 (WiMAX) is an offshoot of IEEE 802, which applies to Ethernet, the technology that powers the Category 5 cable, which connects the vast majority of the world's computers. In Ethernet, the PHY is usually contained in a Category 5 cable. In short, WiMAX and the preceding standard 802.11 (Wi-Fi) are wireless forms of Ethernet. Therefore, much of the Open Systems Interconnection (OSI) Reference Model applies. Figure 2-1 details the OSI Reference Model as it relates to 802.11, and Figure 2-2 outlines the 802.16 PHY and Medium Access Control (MAC) layer.

As they are wireless versions of Ethernet, IEEE standards 802.11 and 802.16 employ a PHY and a MAC layer to accommodate the wireless medium. Figure 2-1 illustrates the IEEE 802.11 variations of the OSI model. Note how the PHY and data link layers have been subdivided to accommodate the wireless medium. Figure 2-2 details MAC and physical layers in 802.16

The Function of the PHY

As the name might imply, the purpose of the PHY is the physical transport of data. The following paragraphs will describe different methods to ensure the most efficient delivery in terms of bandwidth (volume and time in Mbps) and frequency spectrum (MHz/GHz). A number of legacy technologies are used to get the maximum performance out of the PHY. These technologies, including OFDM, TDD, FDD, QAM, and Adaptive Antenna System (AAS), will be described in the following pages or chapters.

OFDM: The "Big So What?!" of WiMAX

OFDM is what puts the *max* in WiMAX. OFDM is not new. Bell Labs originally patented it in 1970, and it became incorporated in various digital subscriber line (DSL) technologies as well as in 802.11a. OFDM is based on a mathematical process called Fast Fourier Transform (FFT), which enables 52 channels to overlap without losing their individual characteristics (orthogonality). This is a more efficient use of the spectrum and enables the channels to be processed at the receiver more efficiently. OFDM is especially popular in wireless applications because of its resistance to forms of interference and degradation (multipath and delay spread, more on this in Chapter 6). In short, OFDM delivers a wireless signal much farther with less interference than competing technologies. Figure 2-3 provides an illustration of how OFDM works.

Figure 2-3
The significance of OFDM: A focused beam delivering maximum bandwidth over maximum distance with minimum interference

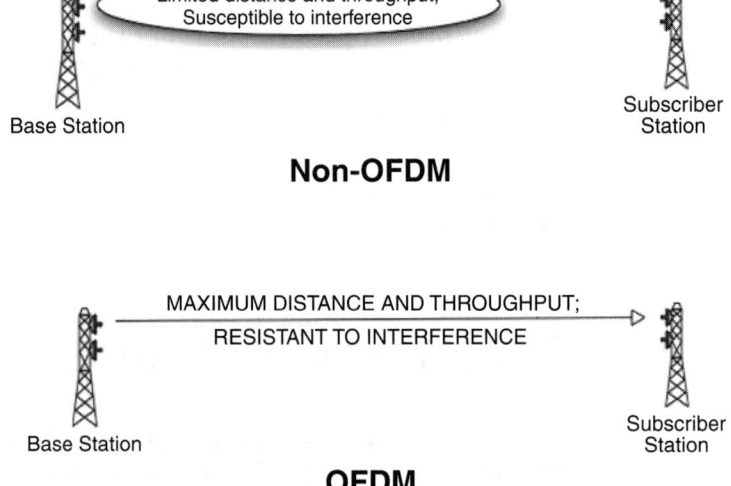

TDD and FDD

WiMAX supports both time division duplex (TDD) and frequency division duplex (FDD) operation. TDD is a technique in which the system transmits and receives within the same frequency channel, assigning time slices for transmit and receive modes. FDD requires two separate frequencies generally separated by 50 to 100 MHz within the operating band. TDD provides an advantage where a regulator allocates the spectrum in an adjacent block. With TDD, band separation is not needed, as is shown in Figure 2-4. Thus, the entire spectrum allocation is used efficiently both upstream and downstream and where traffic patterns are variable or asymmetrical.

In FDD systems, the downlink (DL) and uplink (UL) frame structures are similar except that the DL and UL are transmitted on separate channels. When half-duplex FDD (H-FDD) subscriber stations (SSs) are present, the base station (BS) must ensure that it does not schedule an H-FDD SS to transmit and receive at the same time.[1]

Figure 2-5 illustrates this relationship.

Figure 2-4
A TDD subframe

Frame Header	Downlink Subframe	TG	Uplink Subframe

Figure 2-5
ULs and DLs
between BSs
and SSs

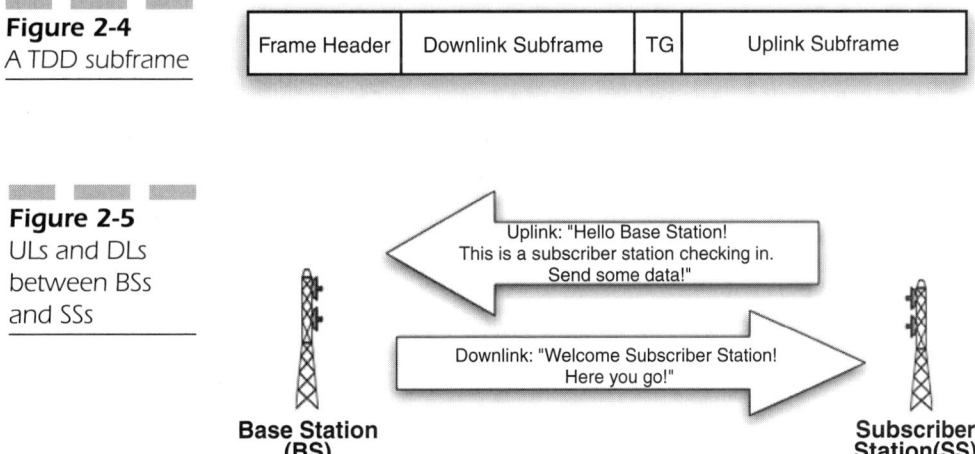

Uplink: "Hello Base Station! This is a subscriber station checking in. Send some data!"

Downlink: "Welcome Subscriber Station! Here you go!"

Base Station (BS)　　　　　**Subscriber Station(SS)**

[1]Govindan Nair, Joey Chou, Tomaz Madejski, Krzysztof Perycz, David Putzolu, and Jerry Sydir, "IEEE 802.16 Medium Access Control and Service Provisioning," *Intel Technology Journal* 3, no. 3 (August 20, 2004): 216–217.

Figure 2-6
AAS uses beam forming to increase gain (energy) to the intended SS.

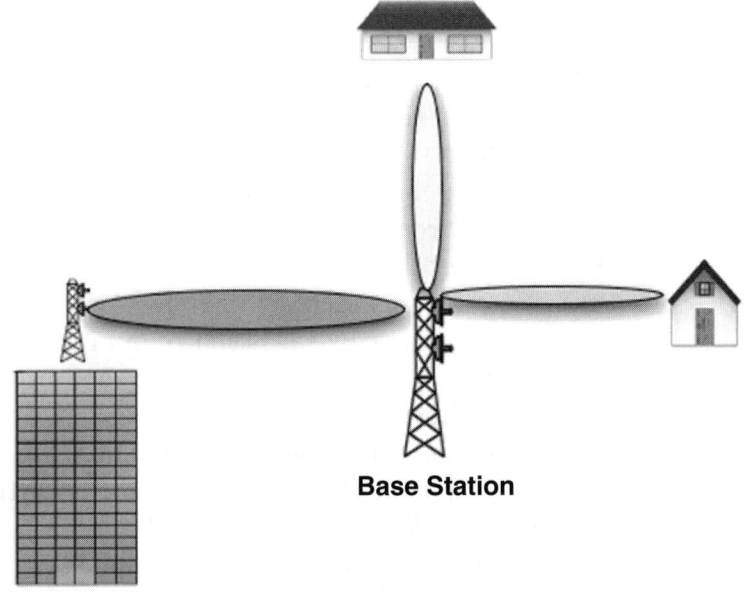

Figure 2-6
AAS uses beam forming to increase gain (energy) to the intended SS.

Base Station

Adaptive Antenna System (AAS)

AAS is used in the WiMAX specification to describe beam-forming techniques where an array of antennas is used at the BS to increase gain to the intended SS while nulling out interference to and from other SSs and interference sources. AAS techniques can be used to enable Spatial Division Multiple Access (SDMA), so multiple SSs that are separated in space can receive and transmit on the same subchannel at the same time. By using beam forming, the BS is able to direct the desired signal to the different SSs and can distinguish between the signals of different SSs, even though they are operating on the same subchannel(s), as shown in Figure 2-6.

WiMAX Variants

WiMAX has five variants, which are specified by their PHY. The variants are divided by whether the variant is single carrier (SC) or uses OFDM. They are further broken down into the frequency bands they cover: 2–11 GHz and 10–66 GHz. The following paragraphs give a brief overview of each variant with emphasis on Wireless metro area network–OFDM (aka WirelessMAN-OFDM). Much of the following is for reference purposes, and the less technical reader may want to move on to Chapter 3 at this time. Table 2-1 provides an overview of these variants.

OFDM Variants 2–11 GHz

The need for NLOS operation drives the design of the 2–11 GHz PHY. Because residential applications are expected, rooftops may be too low (possibly due to obstruction by trees or other buildings) for a clear sight line to a BS antenna. Therefore, significant multipath propagation must be expected. Furthermore, outdoor-mounted antennas are expensive, due to both hardware and installation costs. The four 2–11 GHz air interface specifications are described in the following paragraphs.

WirelessMAN-OFDM   This air interface uses OFDM with a 256-point transform (see OFDM description later in this chapter). Access is by TDMA. This air interface is mandatory for license-exempt bands.

Table 2-1

Variants of WiMAX PHY

Designation	Function	LOS/ NLOS	Frequency	Duplexing Alternative(s)
WirelessMAN-SC	Point-to-point	LOS	10–66 GHz	TDD, FDD
WirelessMAN-SCa	Point-to-point	NLOS	2–11 GHz	TDD FDD
WirelessMAN OFDM	Point-to-mulitpoint	NLOS	2–11 GHz	TDD FDD
WirelessMAN-OFDMA	Point-to-mulitpoint	NLOS	2–11 GHz	TDD FDD
Wireless HUMAN	Point-to-mulitpoint	NLOS	2–11 GHz	TDD

The WirelessMAN-OFDM PHY is based on OFDM modulation. It is intended mainly for fixed access deployments where SSs are residential gateways deployed within homes and businesses. The OFDM PHY supports subchannelization in the UL. There are 16 subchannels in the UL. The OFDM PHY supports TDD and FDD operations, with support for both FDD and H-FDD SSs. The standard supports multiple modulation levels including Binary Phase Shift Keying (BPSK), QPSK, 16-QAM, and 64-QAM. Finally, the PHY supports (as options) transmit diversity in the DL using Space Time Coding (STC) and AAS with Spatial Division Multiple Access (SDMA).

The transmit diversity scheme uses two antennas at the BS to transmit an STC-encoded signal to provide the gains that result from second-order diversity. Each of two antennas transmits a different symbol (two different symbols) in the first symbol time. The two antennas then transmit the complex conjugate of the same two symbols in the second symbol time. The resulting data rate is the same as without transmit diversity.

Figure 2-7 illustrates the frame structure for a TDD system. The frame is divided into DL and UL subframes. The DL subframe is made up of a preamble, Frame Control Header (FCH), and a number of data bursts. The FCH specifies the burst profile and the length of one or more DL bursts that immediately follow the FCH. The down-

Figure 2-7
Frame structure
for a TDD
system
(Source: IEEE)

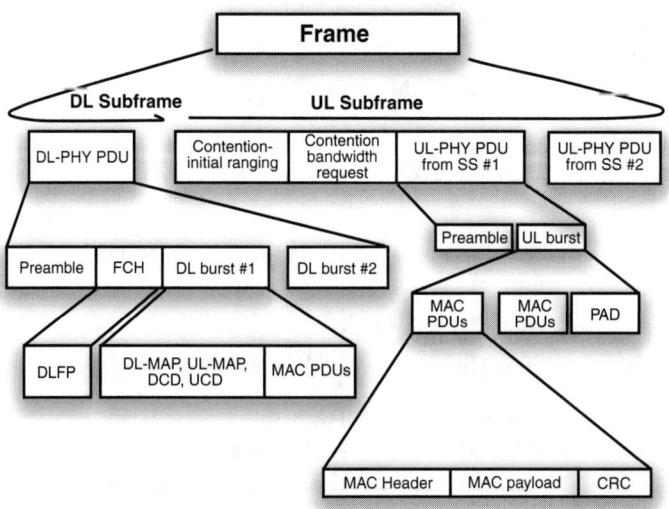

link map (DL-MAP), uplink map (UL-MAP), DL Channel Descriptor (DCD), UL Channel Descriptor (UCD), and other broadcast messages that describe the content of the frame are sent at the beginning of these first bursts. The remainder of the DL subframe is made up of data bursts to individual SSs.

Each data burst consists of an integer number of OFDM symbols and is assigned a burst profile that specifies the code algorithm, code rate, and modulation level that are used for those data transmitted within the burst. The UL subframe contains a contention interval for initial ranging and bandwidth allocation purposes and UL PHY protocol data units (PDUs) from different SSs. The DL-MAP and UL-MAP completely describe the contents of the DL and UL subframes. They specify the SSs that are receiving and/or transmitting in each burst, the subchannels on which each SS is transmitting (in the UL), and the coding and modulation used in each burst and in each subchannel.

If transmit diversity is used, a portion of the DL frame (called a zone) can be designated to be a transmit diversity zone. All data bursts within the transmit diversity zone are transmitted using STC coding. Finally, if AAS is used, a portion of the DL subframe can be designated as the AAS zone. Within this part of the subframe, AAS is used to communicate to AAS-capable SSs. AAS is also supported in the UL.

WirelessMAN-OFDMA This variant uses orthogonal frequency division multiple access (OFDMA) with a 2048-point transform (a function of OFDM, see Chapter 5 for a description). In this system, addressing a subset of the multiple carriers to individual receivers provides multiple access. Because of the propagation requirements, the use of AASs is supported.

The WirelessMAN-OFDMA PHY is based on OFDM modulation. It supports subchannelization in both the UL and DL. The standard supports five different subchannelization schemes. The OFDMA PHY supports both TDD and FDD operations. The same modulation levels are also supported. STC and AAS with SDMA are supported, as is multiple input, multiple output (MIMO). MIMO encompasses a number of techniques for utilizing multiple antennas at the BS and SS in order to increase the capacity and range of the channel.

The frame structure in the OFDMA PHY is similar to the structure of the OFDM PHY. The notable exceptions are that subchannelization is defined in the DL as well as in the UL, so broadcast messages are sometimes transmitted at the same time (on different subchannels) as data. Also, because a number of different subchannelization schemes are defined, the frame is divided into a number of zones that each use a different subchannelization scheme. The MAC layer is responsible for dividing the frame into zones and communicating this structure to the SSs in the DL-MAP and UL-MAP. As in the OFDM PHY, there are optional transmit diversity and AAS zones, as well as a MIMO zone.[2]

Wireless High Speed Unlicensed Metro Area Network (WirelessHUMAN)   WirelessHUMAN is similar to the aforementioned OFDM-based schemes and is focused on Unlicensed National Information Infrastructure (UNII) devices and other unlicensed bands.

Single Carrier (SC) Variants

There are two single carrier variants of WiMAX. These variants are founded on frequency division duplexing and time division duplexing.

WirelessMan-SC 10–66 GHz   In this point-to-multipoint architecture, the BS basically transmits a time division multiplexing (TDM) signal, with individual SS allocated time slots serially. WirelessMAN-SC 10–66 GHz utilizes a burst design that allows both TDD, in which the UL and DL share a channel but do not transmit simultaneously, and FDD, in which the UL and DL sometimes operate simultaneously on separate channels. This burst design allows both TDD and FDD to be handled similarly. Moreover, both TDD and FDD alternatives support adaptive burst profiles in which modulation and coding options may be dynamically assigned on a burst-by-burst basis. Chapter 5 describes this procedure in greater detail.

---

[2]Ibid., 216.

Uplinks (ULs) The UL in the PHY is based on a combination of TDMA and demand assigned multiple access (DAMA). The UL channel is divided into a number of time slots. The MAC layer in the BS controls the number of slots (which may vary over time for optimal performance) assigned for various uses (registration, contention, guard, or user traffic). The DL channel is TDM, with the information for each SS multiplexed onto a single stream of data and received by all SSs within the same sector. To support H-FDD SSs, provision is also made for a TDMA portion of the DL.[3]

A typical UL subframe for the 10–66 GHz PHY is shown in Figure 2-8. Unlike the DL, the UL-MAP grants bandwidth to specific SSs. The SSs transmit in their assigned allocation using the burst profile specified by the Uplink Interval Usage Code (UIUC) in the UL-MAP entry granting them bandwidth. The UL subframe may also contain contention-based allocations for initial system access and broadcast or multicast bandwidth requests. The access opportunities for initial system access are sized to allow extra guard time for SSs that have not resolved the transmit time advance necessary to offset the round trip delay to the BS.[4]

Downlinks (DLs) The DL PHY includes a Transmission Convergence sublayer that inserts a pointer byte at the beginning of the payload to help the receiver identify the beginning of a MAC PDU. Data bits coming from the Transmission Convergence sublayer are

Figure 2-8
UL subframe for
WirelessMAN-SC
(Source: IEEE)

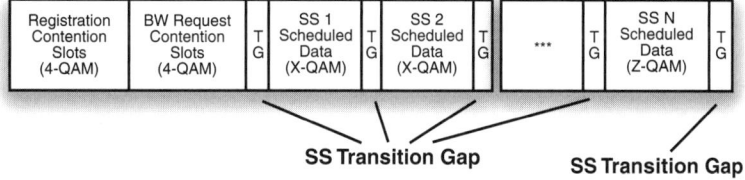

---

[3]"802.16-2004 IEEE Standard for Local and Metropolitan Area Networks, Part 16, Air Interface for Fixed Broadband Wireless Access Systems," June 24, 2004, 307.

[4]Roger Marks, Carl Eklund, Kenneth Stanwood, and Stanley Wang, "IEEE 802.16: A Technical Overview of the WirelessMAN Air Interface for Broadband Wireless Access," *IEEE Communications*, June 2002, 100–101.

randomized, forward error correction (FEC) encoded, and mapped to a QPSK, 16-QAM, or 64-QAM (optional) signal constellation.[5] (Modulation schemes will be covered in detail in Chapter 5.) In the structure for a burst FDD DL frame, each frame is subdivided into a number of physical slots, and each slot represents four modulation symbols. The frame starts with a TDM section that is organized into different modulation and FEC groups. The groups contain data transmitted to full-duplex stations. The last section of the frame is the TDMA section, which contains data transmitted to the half-duplex stations.

Each burst upstream frame contains three types of slots: (1) contention slots used for registration, (2) contention slots used for bandwidth/channel requests, and (3) slots reserved for individual stations. Each type of slot carries the modulation scheme that it is supposed to support, and different stations can be assigned different modulation schemes. The contention slots use 4-QAM, but the reserved slots can be assigned any modulation scheme.

In continuous FDD, the upstream channel is partitioned into a series of minislots, and each minislot consists of a group of physical slots. As stated earlier, a physical slot consists of four modulation symbols. The BS periodically broadcasts the upstream MAP message on the downstream channel. The upstream MAP message defines the permissible usage of each upstream minislot within the time interval covered by the MAP message. Upstream MAP messages are transmitted approximately 250 times per second. This is illustrated in Figure 2-9.

Figure 2-9
TC sublayer
and the MAC
PDU in the
WirelessMAN-
SC PHY
(Source: IEEE)

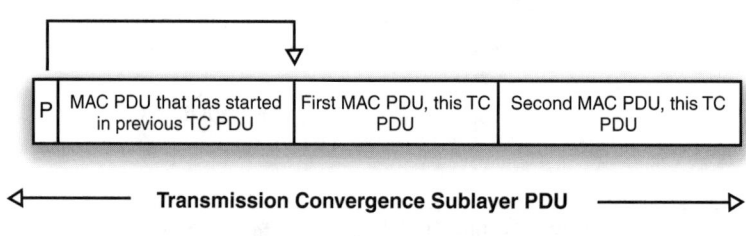

The FEC used in WiMAX is Reed-Solomon GF(256), with variable block size and error correction capabilities. This is paired with an inner block convolutional code to robustly transmit critical data such as frame control and initial accesses. The FEC options are paired with QPSK, 16-QAM, and 64-QAM to form burst profiles of varying robustness and efficiency. If the last FEC block is not filled, that block may be shortened. Shortening in both the UL and DL is controlled by the BS and is communicated in the UL-MAP and DL-MAP.

The system uses a frame of 0.5, 1, or 2 ms. This frame is divided into physical slots for the purpose of bandwidth allocation and identification of PHY transitions. A physical slot is defined to be four QAM symbols. In the TDD variant of the PHY, the UL subframe follows the DL subframe on the same carrier frequency. In the FDD variant, the UL and DL subframes are coincident in time but are carried on separate frequencies. The DL subframe is shown in Figure 2-10.

DL Subframe   The DL subframe starts with a frame control section that contains the DL-MAP for the current DL frame as well as the UL-MAP for a specified time in the future. The DL-MAP specifies when PHY transitions (modulation and FEC changes) occur

Figure 2-10
FDD DL
subframe
(Source: IEEE)

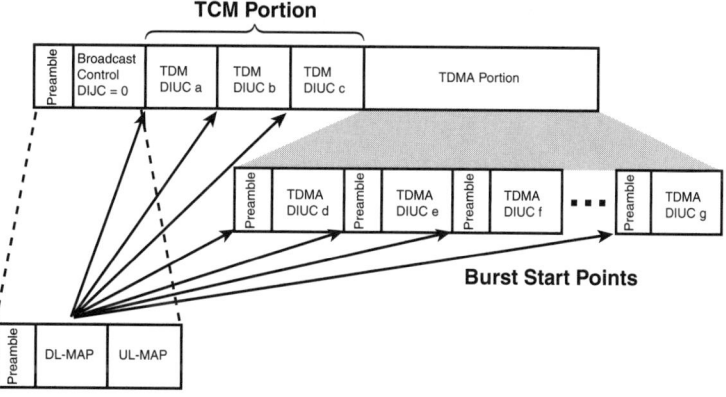

FDD Downlink Subframe

within the DL subframe. The DL subframe typically contains a TDM portion immediately following the frame control section. DL data are transmitted to each SS using a negotiated burst profile. The data are transmitted in order of decreasing robustness to allow SSs to receive their data before being presented with a burst profile that could cause them to lose synchronization with the DL.

In FDD systems, a TDMA segment that includes an extra preamble at the start of each new burst profile may follow the TDM portion. This feature allows better support of half-duplex SSs. In an efficiently scheduled FDD system with many half-duplex SSs, some SSs may need to transmit earlier in the frame than they receive. Due to their half-duplex nature, these SSs lose synchronization with the DL. The TDMA preamble allows them to regain synchronization.

Due to the dynamics of bandwidth demand for the variety of services that may be active, the mixture and duration of burst profiles and the presence or absence of a TDMA portion vary dynamically from frame to frame. Because the recipient SS is implicitly indicated in the MAC headers rather than in the DL-MAP, SSs listen to all portions of the DL subframe they are capable of receiving. For full-duplex SSs, this means receiving all burst profiles of equal or greater robustness than they have negotiated with the BS.

WirelessMAN–Single Carrier Access (WirelessMAN-SCa) 2–11 GHz This variant uses a single-carrier modulation format in the 2–11 GHz spectrum and is designed for NLOS operations. Five concepts define the WirelessMAN-SCa variant of the PHY. Elements within this PHY include TDD and FDD definitions (one of which must be supported), TDMA UL, TDM or TDMA DL, and block adaptive modulation. The PHY also includes FEC coding for both UL and DL and framing structures that enable improved equalization, channel estimation performance over NLOS and extended delay spread environments, parameter settings, and MAC/PHY messages that facilitate optional AAS implementations.[6] Table 2-2 further defines elements in this sub specification.

---

[6]Ibid.

Table 2-2

Components Contained in WirelessMAN-SCa 2–11 GHz

Term	Description
Payload	Payload refers to individual units of transmission content that are of interest to some entity at the receiver end.
Burst	A burst contains payload data and is formed according to the rules specified by the burst profile associated with the burst. The existence of the burst is made known to the receiver through the contents of either the UL-MAP or DL-MAP. For the UL, a burst is a complete unit of transmission that includes a leading preamble, encoded payload, and trailing termination sequence.
Burst Set	A burst set is a self-contained transmission entity consisting of a preamble, one or more concatenated bursts, and a trailing termination sequence. For the UL, burst set is synonymous with burst.
Burst Frame	A burst frame contains all information included in a single transmission. It consists of one or more burst sets. The DL and UL subframes each hold a burst frame.
MAC Frame	A MAC frame refers to the fixed bandwidth intervals reserved for data exchange. For TDD, a MAC frame consists of one DL and one UL subframe, delimited by the TTG. For FDD, the MAC frame corresponds to the maximum length of the DL subframe. FDD UL subframes operate concurrently with DL subframes but on a separate (frequency) channel.

Conclusion

If there were one word to describe the WiMAX PHY, it would be *robust*. That is, it uses tested legacy technologies to deliver maximum bandwidth over maximum distances with minimum loss to interference. Because multiple variants of the PHY have been built into the specification, the standard can be applied to multiple roles within a wireless network. For example, the SC variant is well suited for point-to-point backhaul applications, and the OFDM variant is well suited for last-mile point-to-multipoint applications. Together, these variants and their underlying technologies are the building blocks for a next generation broadband wireless network.

The Medium Access Control (MAC) Layer

The MAC as the "Smarts" for the Physical Layer

The WiMAX MAC provides intelligence for the PHY and ensures a number of QoS measures not seen on other wireless standards. Perhaps its greatest value is providing for dynamic bandwidth allocation that defeats the usual degradations of wireless services—jitter and latency.

The WiMAX MAC protocol was designed for point-to-multipoint broadband wireless access applications. It addresses the need for very high bit rates, both UL (to the BS) and DL (from the BS). With WiMAX, unlike with its Wi-Fi predecessors, access and bandwidth allocation algorithms accommodate hundreds of terminals per channel, and multiple end users might share those terminals. End users require services that are varied in nature including legacy TDM voice and data, IP connectivity, and packetized VoIP. To support these various services, the WiMAX MAC accommodates both continuous and bursty traffic. Additionally, these services expect to be assigned QoS parameters in keeping with the traffic types.

The WiMAX MAC protocol supports a variety of backhaul requirements including both ATM and packet-based protocols. Convergence sublayers map the transport-layer-specific traffic to a MAC that is flexible enough to efficiently carry any traffic type. The convergence sublayers and MAC work together using payload header suppression, packing, and fragmentation to carry traffic more efficiently than the original transport mechanism.

The MAC and WiMAX Architecture

The WiMAX DL from the BS to the user operates on a point-to-multipoint basis as illustrated in Figure 3-1. The WiMAX wireless link operates with a central BS with a sectorized antenna that is capable of handling multiple independent sectors simultaneously. Within a given frequency channel and antenna sector, all stations receive the same transmission. The BS is the only transmitter operating in this direction, so it transmits without having to coordinate with other stations except the overall TDD that may divide time into UL and DL transmission periods. The DL is generally broadcast. In cases where the DL-MAP does not explicitly indicate that a portion of the DL subframe is not a specific SS, all SSs capable of listening to that portion of the DL subframe will listen.

The MAC is connection-oriented. Connections are referenced with 16-bit connection identifiers (CIDs) and may require continuously granted bandwidth or bandwidth on demand. As described previously, both bandwidths are accommodated. A CID is used to distinguish between multiple UL channels that are associated with the same DL channel. The SSs check the CIDs in the received PDUs and retain only those PDUs addressed to them.

The MAC PDU is the data unit exchanged between the MAC layers of the BS and its SSs. It is the data unit generated on the downward direction for the next lower layer and the data unit received on the upward direction from the previous lower layer.

Each SS has a standard 48-bit MAC address, which serves as an equipment identifier because the primary addresses used during

Figure 3-1
Typical WiMAX architecture for point-to-multipoint distribution

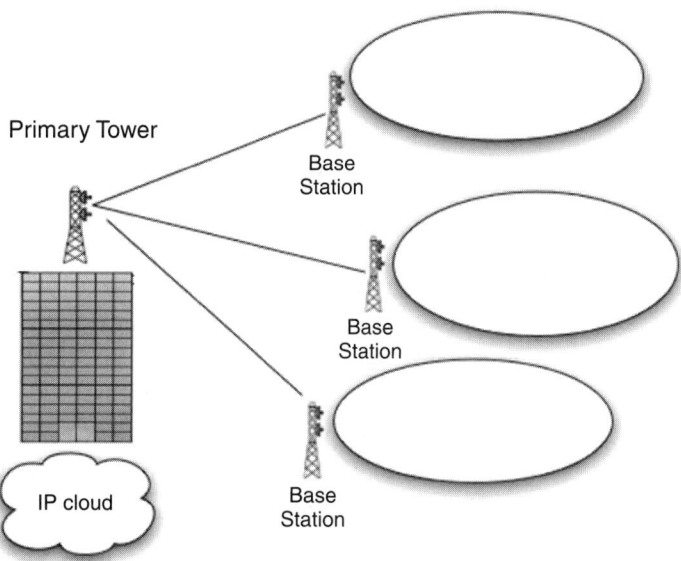

operation are the CIDs. Upon entering the network, the SS is assigned three management connections in each direction. These three connections reflect the three different QoS requirements used by different management levels:

- **Basic connection**—transfers short, time-critical MAC and radio link control (RLC) messages (see Chapter 4).

- **Primary management connection**—transfers longer, more delay-tolerant messages, such as those used for authentication and connection setup. The secondary management connection transfers standards-based management messages such as Dynamic Host Configuration Protocol (DHCP), Trivial File Transfer Protocol (TFTP), and Simple Network Management Protocol (SNMP). In addition to these management connections, SSs are allocated transport connections for the contracted services.

- **Transport connections**—are unidirectional to facilitate different UL and DL QoS and traffic parameters; they are typically assigned to services in pairs.

SSs share the UL to the BS on a demand basis. Depending on the class of service utilized, the SS may be issued continuing rights to transmit, or the BS may grant the right to transmit after receiving a request from the user.

Service Classes and QoS

Within each sector, users adhere to a transmission protocol that controls contention between users and enables the service to be tailored to the delay and bandwidth requirements of each user application. This is accomplished through four different types of UL scheduling mechanisms. These mechanisms are implemented using unsolicited bandwidth grants, polling, and contention procedures. The WiMAX MAC provides QoS differentiation for different types of applications that might operate over WiMAX networks:

- **Unsolicited Grant Services (UGS)**—UGS is designed to support constant bit rate (CBR) services, such as T1/E1 emulation and VoIP without silence suppression.

- **Real-Time Polling Services (rtPS)**—rtPS is designed to support real-time services that generate variable size data packets, such as MPEG video or VoIP with silence suppression, on a periodic basis.

- **Non-Real-Time Polling Services (nrtPS)**—nrtPS is designed to support non-real-time services that require variable size data grant burst types on a regular basis.

- **Best Effort (BE) Services**—BE services are typically provided by the Internet today for web surfing.

The use of polling simplifies the access operation and guarantees that applications receive service on a deterministic basis if required. In general, data applications are delay tolerant, but real-time applications, like voice and video, require service on a more uniform basis and sometimes on a very tightly controlled schedule.

For the purposes of mapping to services on SSs and associating varying levels of QoS, all data communications are in the context of a connection. Service flows may be provisioned when an SS is installed in the system. Shortly after SS registration, connections are associated with these service flows (one connection per service flow) to provide a reference against which to request bandwidth. Additionally, new connections may be established when a customer's service needs change. A connection defines both a service flow and the mapping between peer convergence processes that utilize the MAC. The service flow defines the QoS parameters for the PDUs that are exchanged once the connection has been established.

Service flows are the mechanism for UL and DL for QoS management. In particular, they facilitate the bandwidth allocation process. An SS requests UL bandwidth on a per connection basis (implicitly identifying the service flow). The BS grants the bandwidth to an SS as an aggregate of grants in response to per connection requests from the SS.[1]

The modulation and coding schemes are specified in a burst profile that may be adjusted adaptively for each burst to each SS. The

---

[1]"802.16-2004 IEEE Standard for Local and Metropolitan Area Networks, Part 16, Air Interface for Fixed Broadband Wireless Access Systems," June 24, 2004, 31.

MAC can make use of bandwidth-efficient burst profiles under favorable link conditions then shift to more reliable, although less efficient alternatives, as required to support the planned 99.999 percent link availability (QPSK to 16-QAM to 64-QAM).

The request-grant mechanism is designed to be scalable, efficient, and self-correcting. The WiMAX access system does not lose efficiency when presented with multiple connections per terminal, multiple QoS levels per terminal, and a large number of statistically multiplexed users.

Along with the fundamental task of allocating bandwidth and transporting data, the MAC includes a privacy sublayer that provides authentication of network access and connection establishment to avoid theft of service, and it provides key exchange and encryption for data privacy.

Service-Specific Convergence Sublayers

The WiMAX standard defines two general service-specific convergence sublayers for mapping services to and from WiMAX MAC connections:

- The ATM convergence sublayer is for ATM services.
- The packet convergence sublayer is defined for mapping packet services such as Internet Protocol version 4 or 6 (IPv4, IPv6), Ethernet, and virtual local area network (VLAN).

The primary task of the sublayer is to classify service data units (SDUs) to the proper MAC connection, preserve or enable QoS, and enable bandwidth allocation. SDUs are the units exchanged between two adjacent protocol layers. They are the data units received on the downward direction from the previous higher layer and the data units sent on the upward direction to the next higher layer. The mapping takes various forms, depending on the type of service. In addition to these basic functions, the convergence sublayers perform sophisticated functions, such as payload header suppression and reconstruction, to enhance airlink efficiency.

Common Part Sublayer

The MAC reserves additional connections for other purposes. One connection is reserved for contention-based initial access. Another is reserved for broadcast transmissions in the DL as well as for signaling broadcast contention-based polling of SS bandwidth needs. Additional connections are reserved for multicast, rather than broadcast, contention-based polling. SSs may be instructed to join multicast polling groups associated with these multicast polling connections.

MAC PDU Formats   A MAC PDU consists of a fixed-length MAC header, a variable-length payload, and an optional cyclic redundancy check (CRC). Two header formats are defined: the generic header (as illustrated in Figure 3-2) and the bandwidth request header. Except for bandwidth request MAC PDUs, which contain no payload, MAC PDUs contain either MAC management messages or convergence sublayer data.

There are three types of MAC subheaders:

- **Grant management subheader**—is used by an SS to convey bandwidth management needs to its BS.

- **Fragmentation subheader**—contains information that indicates the presence and orientation in the payload of any fragments of SDUs.

- **Packing subheader**—indicates the packing of multiple SDUs into a single PDU. The grant management and fragmentation subheaders may be inserted in MAC PDUs immediately following the generic header if so indicated by the Type field. The packing subheader may be inserted before each MAC SDU if so indicated by the Type field.

Figure 3-2
MAC PDU
(Source: IEEE)

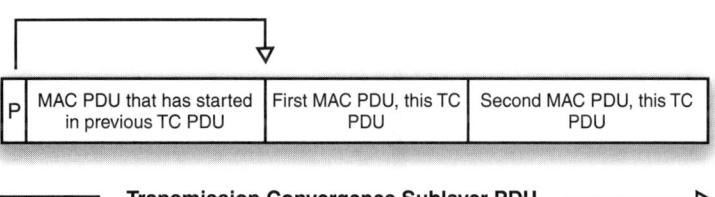

Transmission of MAC PDUs and SDUs   Incoming MAC SDUs from corresponding convergence sublayers are formatted according to the MAC PDU format, with fragmentation and/or packing, before being conveyed over one or more connections in accordance with the MAC protocol. After traversing the airlink, MAC PDUs are reconstructed into the original MAC SDUs so that the format modifications performed by the MAC layer protocol are transparent to the receiving entity. This is illustrated in Figure 3-3.

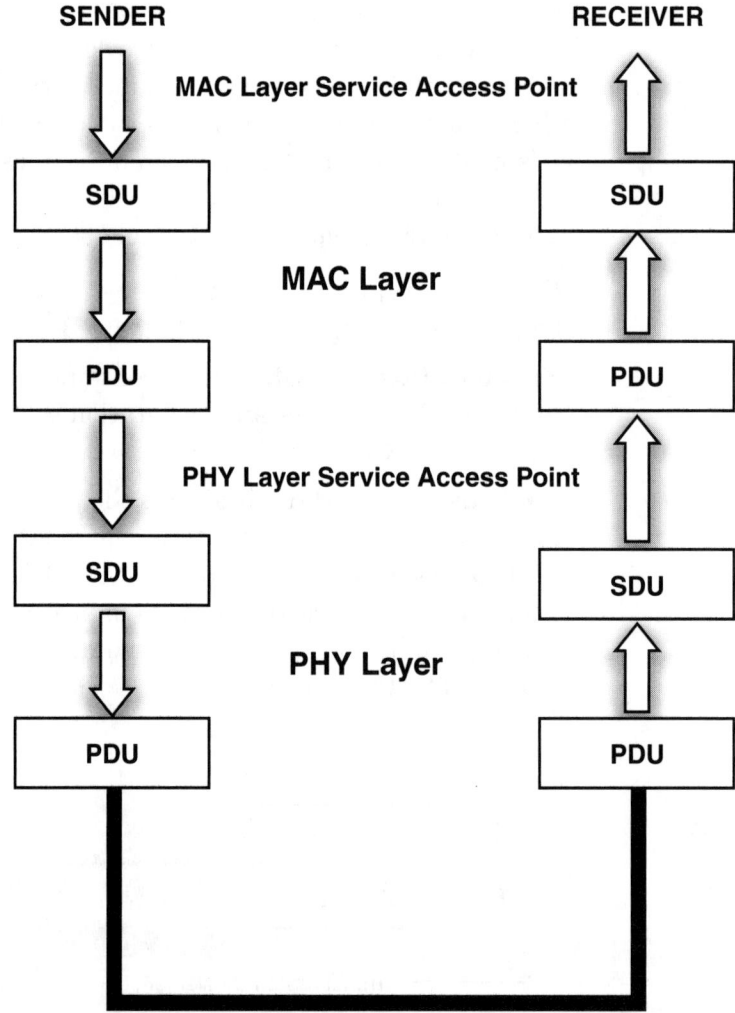

Figure 3-3
Fragmentation and packing of SDUs and PDUs (Source: IEEE)

Packing and Fragmentation

WiMAX takes advantage of incorporating the packing and fragmentation processes with the bandwidth allocation process to maximize the flexibility, efficiency, and effectiveness of both. Fragmentation is the process in which a MAC SDU is divided into one or more MAC SDU fragments. Packing is the process in which multiple MAC SDUs are packed into a single MAC PDU payload. Either a BS for a DL connection or an SS for an UL connection may initiate both processes. WiMAX allows simultaneous fragmentation and packing for efficient use of the bandwidth.

PDU Creation and Automatic Repeat Request (ARQ)

ARQ blocks are distinct units of data that are carried on ARQ-enabled connections. ARQ processing retransmits MAC SDU blocks (aka ARQ blocks) that have been lost or garbled. The WiMAX MAC uses a simple sliding window-based approach where the transmitter can send up to a negotiated number of blocks without receiving an acknowledgment. The receiver sends acknowledgment or negative acknowledgment messages to indicate to the transmitter which SDU blocks have been received and which have been lost. The transmitter retransmits blocks that were lost and moves the sliding window forward when SDU blocks are acknowledged to have been received.

Each SS to BS connection is assigned a service class, as part of the creation of the connection. When packets are classified in the convergence sublayer, the connection into which they are placed is chosen based on the type of QoS guarantees that the application requires.

Figure 3-3 depicts the WiMAX QoS mechanism in supporting multimedia services including TDM voice, VoIP, video streaming, TFTP, hypertext transfer protocol (HTTP), and e-mail.[2]

---

[2]Govindan Nair, Joey Chou, Tomasz Madejski, Krzysztof Perycz, David Putzolu, and Jerry Sydir, "IEEE 802.16 Medium Access Control and Service Provisioning," *Intel Technology Journal* 3, no. 3 (August 20, 2004): 214–215.

PHY Level Support and Frame Structure   The WiMAX MAC supports both TDD and FDD. In FDD, both continuous and burst DLs are supported. Continuous DLs allow for certain robustness enhancement techniques, such as interleaving. Burst DLs (either FDD or TDD) allow the use of more advanced robustness and capacity enhancement techniques, such as subscriber-level adaptive burst profiling and AASs.

The MAC builds the DL subframe starting with a frame control section containing the DL-MAP and UL-MAP messages. These indicate PHY level transitions on the DL as well as bandwidth allocations and burst profiles on the UL.

The DL-MAP is always applicable to the current frame and is always at least two FEC blocks long. The first PHY level transition is expressed in the first FEC block to allow adequate processing time. In both TDD and FDD systems, the UL-MAP provides allocations starting no later than the next DL frame. The UL-MAP can, however, allocate starting in the current frame as long as processing times and round-trip delays are observed. The minimum time between receipt and applicability of the UL-MAP for an FDD system is shown in Figure 3-4.[3]

Figure 3-4
Uplink subframe
(Source: IEEE)

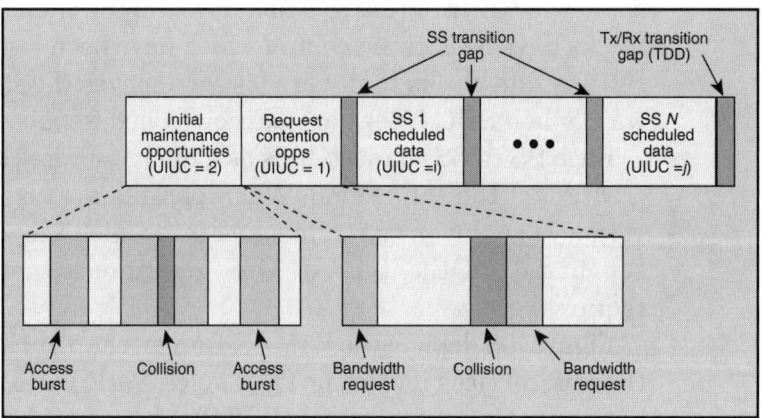

[3]Roger Marks, Carl Eklund, Kenneth Stanwood, and Stanley Wang, "IEEE 802.16: A Technical Overview of the WirelessMAN Air Interface for Broadband Wireless Access," *IEEE Communications*, June 2002, 102–103.

Transmission Convergence (TC) Layer

Between the PHY and MAC is a TC sublayer (see Figure 3-5). This layer transforms variable length MAC PDUs into fixed-length FEC blocks (plus possibly a shortened block at the end of each burst). The TC layer has a PDU sized to fit in the FEC block currently being filled. It starts with a pointer indicating where the next MAC PDU header starts within the FEC block. This was shown in Figure 3-3. The TC PDU format allows resynchronization to the next MAC PDU in the event that the previous FEC block had irrecoverable errors.

Figure 3-5
Relationship of transmission convergence layer with physical and MAC layers (Source: IEEE)

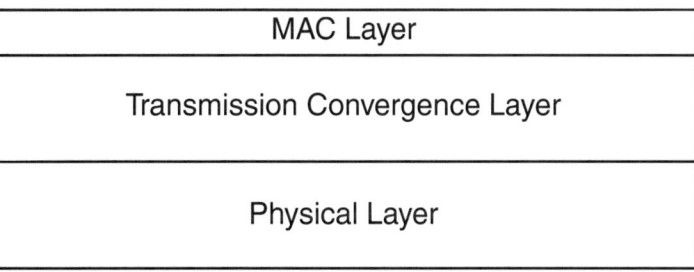

MAC Layer

Transmission Convergence Layer

Physical Layer

How WiMAX
Works

Like most data communications, WiMAX relies on a process consisting of a session setup and authentication. The RLC manages and monitors the quality of the service flow. With WiMAX, this process is a series of exchanges (DLs and ULs) between the BS and SS. A complex process determines what FDD and TDD settings will be used for the service flow, FEC, sets encryption, bandwidth requests, burst profiles, and so on. The process starts with channel acquisition by the newly installed SS.

Channel Acquisition

The MAC protocol includes an initialization procedure designed to eliminate the need for manual configuration. In other words, the subscriber takes the SS out of the box, plugs in power and Ethernet, and connects almost immediately to the network. The following paragraphs describe how that is possible without laborious user setup or service provider truck roll.

Upon installation, the SS begins scanning its frequency list to find an operating channel. It may be preconfigured by the service provider to register with a specified BS. This feature is useful in dense deployments where the SS might hear a secondary BS due to spurious signals or when the SS picks up a sidelobe of a nearby BS antenna. Moreover, this feature will help service providers avoid expensive installations and subsequent truck rolls.

After selecting a channel or channel pair, the SS synchronizes to the DL transmission from the BS by detecting the periodic frame preambles. Once the PHY is synchronized, the SS will look for the periodically broadcasted DCD and UCD messages that enable the SS to determine the modulation and FEC schemes used on the BS's carrier.

Initial Ranging and Negotiation of SS Capabilities

Once the parameters for initial ranging transmissions are established, the SS will scan the UL-MAP messages present in every frame for ranging information. The SS uses a backoff algorithm to determine which initial ranging slot it will use to send a ranging request (RNG-REQ) message. The SS will then send its burst using the minimum power setting and will repeat with increasingly higher transmission power until it receives a ranging response.

Based on the arrival time of the initial RNG-REQ and the measured power of the signal, the BS adjusts the timing advance and power to the SS with the ranging response (RNG-RSP). The response provides the SS with the basic and primary management CIDs. Once the timing advance of the SS transmissions has been correctly determined, the ranging procedure for fine-tuning the power is done via a series of invited transmissions.

WiMAX transmissions are made using the most robust burst profile. To save bandwidth, the SS next reports its PHY capabilities, including which modulation and coding schemes (see Chapter 2) it supports and whether, in an FDD system, it is half-duplex or full-duplex. The BS, in its response, can deny the use of any capability reported by the SS. See Figure 4-1 for an illustration of this process.

It should be noted here how complex this setup procedure is. The purpose thus far is to ensure a high quality connection between the SS and the BS.

Figure 4-1
Channel acquisition process between an SS and BS

Base Station

Channel Acquisition, Ranging, and Negotiation of Subscriber Station Capabilities

Subscriber Station

1. SS begins scanning presets frequency for base station.

2. BS responds. Synchronizes with SS.

3. Ranging parameters sets UL-MAP messages in every frame. SS bursts with increasing power until it reaches/receives a ranging reponse from BS.

4. BS responds with timing and power adjustments, management CIDs.

5. SS reports its physical layer capabilities (modulation/coding schemes).

6. BS accepts SS; is ready for service flow.

SS Authentication and Registration

Wi-Fi has been dogged with a reputation for lax security. Perhaps the best "horror story" deals with a computer retailer who installed a wireless LAN. A customer purchased a Wi-Fi equipped laptop and, anxious to enjoy it, powered it up in the parking lot of the retailer. The new laptop owner was immediately able to tap into the retailer's Wi-Fi network and was able to capture some customer credit card information. Fortunately, the new laptop owner was a journalist, not a con artist. The story, much to the chagrin of the national retailer and the Wi-Fi industry, made the national news. The Wi-Fi industry has had to work hard to shake the reputation of having loose security measures. A similar story will not easily, if ever, occur with WiMAX.

Each SS contains both a manufacturer-issued factory-installed X.509 digital certificate and the certificate of the manufacturer. The SS in the Authorization Request and Authentication Information messages sends these certificates, which set up the link between the 48-bit MAC address of the SS and its public RSA key, to the BS. The network is able to verify the identity of the SS by checking the certificates and can subsequently check the level of authorization of the SS. If the SS is authorized to join the network, the BS will respond to its request with an authorization reply containing an authorization key (AK) encrypted with the SS's public key and used to secure further transactions.

Upon successful authorization, the SS will register with the network. This will establish the secondary management connection of the SS and determine capabilities related to connection setup and MAC operation. The version of IP used on the secondary management connection is also determined during registration.

IP Connectivity

After registration, the SS attains an IP address via DHCP and establishes the time of day via the Internet Time Protocol. The DHCP server also provides the address of the TFTP server from which the SS can request a configuration file. This file provides a standard interface for providing vendor-specific configuration information. See Figure 4-2 for an illustration of this process.

Connection Setup

Now comes the connection setup, where data (the content) actually flows. WiMAX uses the concept of service flows to define one-way transport of packets on either the DL or the UL. Service flows are characterized by a set of QoS parameters, such as those for latency and jitter. To most efficiently utilize network resources, such as bandwidth and memory, WiMAX adopts a two-phase activation model in which resources assigned to a particular admitted service flow may not be actually committed until the service flow is activated. Each admitted or active service flow is mapped to a MAC connection with a unique CID. In general, service flows in WiMAX are preprovisioned, and the BS initiates the setup of the service flows during SS initialization.

In addition, the BS or the SS can dynamically establish service flows. The SS typically initiates service flows only if there is a

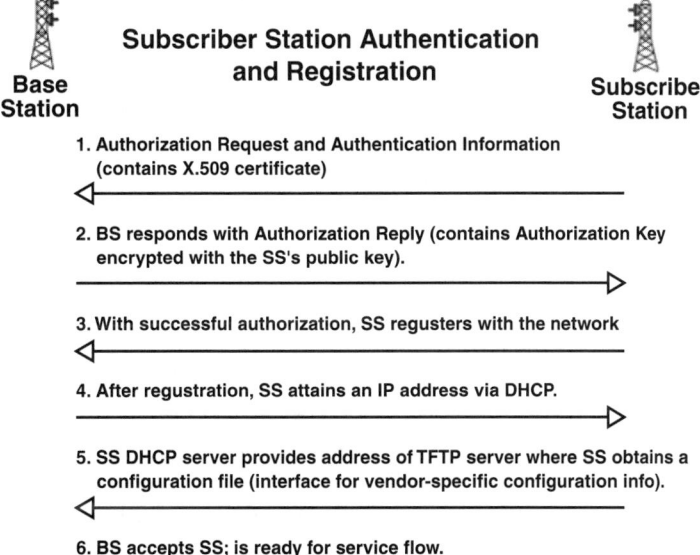

Figure 4-2
SS authentication
and registration

Base Station

Subscriber Station Authentication and Registration

Subscriber Station

1. Authorization Request and Authentication Information (contains X.509 certificate)

2. BS responds with Authorization Reply (contains Authorization Key encrypted with the SS's public key).

3. With successful authorization, SS regusters with the network

4. After regustration, SS attains an IP address via DHCP.

5. SS DHCP server provides address of TFTP server where SS obtains a configuration file (interface for vendor-specific configuration info).

6. BS accepts SS; is ready for service flow.

dynamically signaled connection, such as a switched virtual connection (SVC) from an ATM network. The establishment of service flows is performed via a three-way handshaking protocol in which the request for service flow establishment is responded to and the response acknowledged.

In addition to supporting dynamic service establishment, WiMAX supports dynamic service changes in which service flow parameters are renegotiated. These service flow changes follow a three-way handshaking protocol similar to the one dynamic service flow establishment uses.

Radio Link Control (RLC)

RLC runs simultaneously to channel acquisition and service flow to maintain a steady link. The WiMAX PHY requires equally advanced RLC, particularly the capability of the PHY to transition from one burst profile to another. The RLC controls this capability as well as the traditional RLC functions of power control and ranging.

RLC begins with periodic BS broadcast of the burst profiles that have been chosen for the UP and DL. The particular burst profiles used on a channel are chosen based on a number of factors, such as rain region and equipment capabilities. Burst profiles for the DL are each tagged with a Downlink Interval Usage Code (DIUC). Those for the UL are each tagged with an UIUC.

During initial access, the SS performs initial power leveling and ranging using RNG-REQ messages transmitted in initial maintenance windows. Adjustments to the SS's transmit time advance and power adjustments are returned to the SS in RNG-RSP messages. For ongoing ranging and power adjustments, the BS may transmit unsolicited RNG-RSP messages commanding the SS to adjust its power or timing. This is shown in Figure 4-3.

During initial ranging, the SS also requests to be served in the DL via a particular burst profile by transmitting its choice of DIUC to the BS. The SS performs the choice before and during initial ranging based on received DL signal quality measurements. The BS may confirm or reject the choice in the RNG-RSP. Similarly, the BS monitors the quality of the UL signal it receives from the SS. The BS

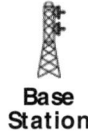

Figure 4-3
RLC ensures
ongoing stability
of the WiMAX
connection

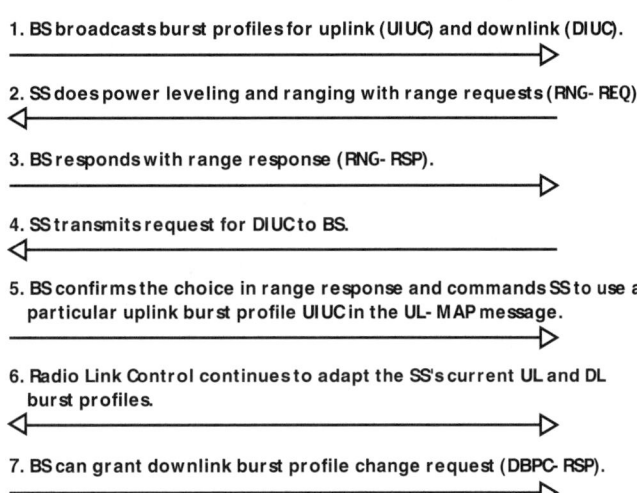

Radio Link Control

**Base
Station**

**Subscriber
Station**

1. BS broadcasts burst profiles for uplink (UIUC) and downlink (DIUC).

2. SS does power leveling and ranging with range requests (RNG-REQ).

3. BS responds with range response (RNG-RSP).

4. SS transmits request for DIUC to BS.

5. BS confirms the choice in range response and commands SS to use a
   particular uplink burst profile UIUC in the UL-MAP message.

6. Radio Link Control continues to adapt the SS's current UL and DL
   burst profiles.

7. BS can grant downlink burst profile change request (DBPC-RSP).

commands the SS to use a particular UL burst profile simply by including the appropriate burst profile UIUC with the SS's grants in UL-MAP messages.

After initially determining UP and DL burst profiles between the BS and a particular SS, RLC continues to monitor and control the burst profiles. Harsher environmental conditions, such as rain fades, can force the SS to request a more robust burst profile. Alternatively, exceptionally good weather may allow an SS to temporarily operate with a more efficient burst profile. The RLC continues to adapt the SS's current UL and DL burst profiles, always striving to achieve a balance between robustness and efficiency.

As the BS controls and directly monitors the UL signal quality, the protocol for changing the UL burst profile for an SS is simple: the BS specifies the profile's UIUC whenever granting the SS bandwidth in a frame. This eliminates the need for an acknowledgment, as the SS will always receive both the UIUC and the grant or neither. This negates the possibility of UL burst profile mismatch between the BS and SS.

In the DL, the SS monitors the quality of the receive signal and knows when to change its DL burst profile. The BS still has ultimate control of the change. The SS has two available methods to request a change in DL burst profile, depending on whether the SS operates in the grant per connection (GPC) or grant per SS (GPSS) mode (see "Bandwidth Requests and Grants" in Chapter 5).

The first method would apply (based on the discretion of the BS scheduling algorithm) only to GPC SSs. In this case, the BS may periodically allocate a station maintenance interval to the SS. The SS can use the RNG-REQ message to request a change in DL burst profile. The preferred method is for the SS to transmit a DL burst profile change request (DBPC-REQ). In this case, which is always an option for GPSS SSs and can be an option for GPC SSs, the BS responds with a DBPC-RSP message confirming or denying the change.

Because messages may be lost due to irrecoverable bit errors, the protocols for changing an SS's DL burst profile must be carefully structured. The order of the burst profile change actions is different when transitioning to a more robust burst profile than when transitioning to a less robust one. The standard takes advantage of the fact that an SS is always required to listen to more robust portions of the DL as well as the profile that was negotiated.[1]

The UL

Each connection in the UL direction is mapped to a *scheduling service*. Each scheduling service is associated with a set of rules imposed on the BS scheduler responsible for allocating the UL capacity and the request-grant protocol between the SS and the BS. The detailed specification of the rules and the scheduling service used for a particular UL connection are negotiated at connection

---

[1]Roger Marks, Carl Eklund, Kenneth Stanwood, and Stanley Wang, "IEEE Standard 802.16: A Technical Overview of the WirelessMAN Air Interface for Broadband Wireless Access," *IEEE Communications*, June 2002, 103–104.

setup time. The scheduling services in WiMAX are based on those defined for cable modems in the Data-Over-Cable Service Interface Specification (DOCSIS) standard.[2]

Service Flow

Minimizing customer intervention and truck roll is very important for WiMAX deployments. The Provisioned Service Flow Table, Service Class Table, and Classifier Rule Table are configured to support self-installation and auto-configuration. When customers subscribe to the service, they tell the service provider the service flow information including the number of UL/DL connections with the data rates and QoS parameters, along with the types of applications (for example, Internet, voice, or video) the customer intends to run. The service provider preprovisions the services by entering the service flow information into the service flow database. When the SS enters the BS by completing the network entry and authentication procedure, the BS downloads the service flow information from the service flow database. Figure 4-4 provides an example of how the service flow information is populated. Figure 4-4A, 4-4B, and 4-4C indicate that two SSs, identified by MAC address 0x123ab54 and 0x45fead1, have been preprovisioned. Each SS has two service flows, identified by sfIndex, with the associated QoS parameters that are identified by qosIndex 1 and 2, respectively. qosIndex points to a QoS entry in the wmanIfBsServiceClassTable that contains three levels of QoS: Gold, Silver, and Bronze. sfIndex points to the entry in the wmanBsClassifierRuleTable and indicates which rules shall be used to classify packets on the given service flow.

When the SS with MAC address 0x123ab54 registers into the BS, the BS creates an entry in the wmanIfBaseRegisteredTable. Based on the MAC address, the BS will be able to find the service flow information that has been preprovisioned. The BS will use a dynamic service activate (DSA) message to create service flows for

---

[2]SCTE DSS 00-05, Data-Over-Cable Service Interface Specification (DOCSIS) SP-RFIv 1-105-000714, "Radio Frequency Interface 1.1 Specification," July 2000.

sfIndex 100001 and 100002, with the preprovisioned service flow information. This can be seen in Figure 4-4. It creates two entries in wmanIfCmnCpsServiceFlowTable. The service flows will then be available for the customer to send data traffic.[3]

Figure 4-4
Service flow provisioning (Source: Intel)

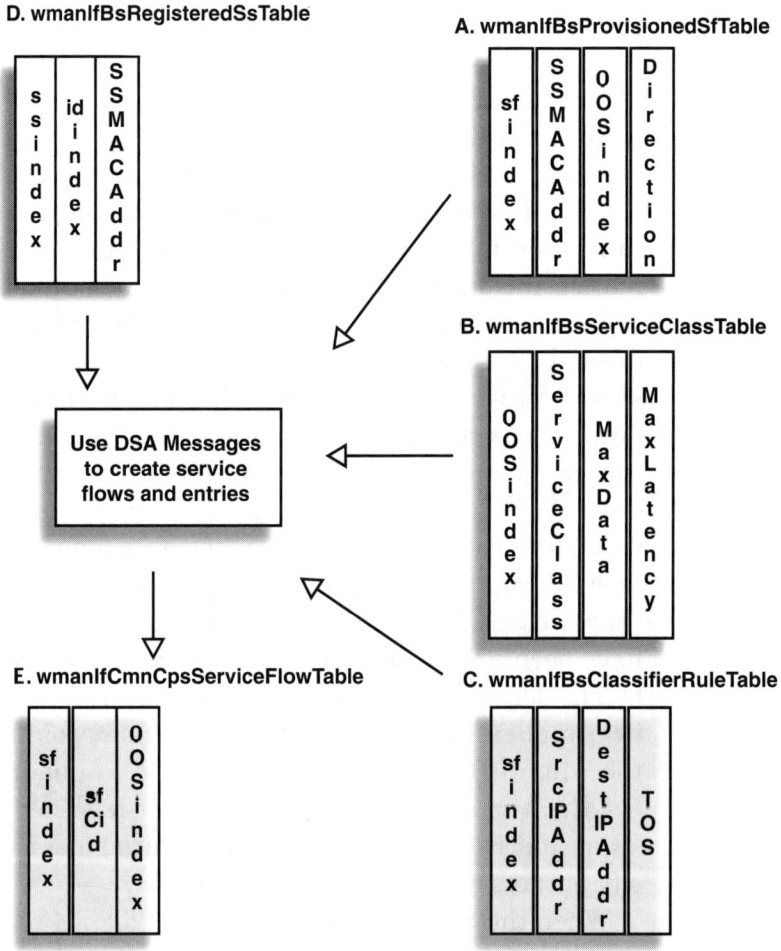

[3]Govindan Nair, Joey Chou, Tomasz Madejski, Krzysztof Perycz, David Putzolu, and Jerry Sydir, "IEEE 802.16 Medium Access Control and Service Provisioning," *Intel Technology Journal* 3, no. 3 (August 20, 2004).

Conclusion

This chapter explains the steps preceding the setup for the WiMAX service flow. The process begins with ranging and negotiation between the BS and SS followed by authentication and registration. This design is noted for its robust nature. The WiMAX RLC then establishes the UL, which sets up the service flow. WiMAX basis in DOCSIS is evident in the sturdy design of this process.

Quality of Service (QoS) on WiMAX

Overview

Perhaps networking *cognoscenti's* greatest objection to broadband wireless access systems is the notion that any data communications protocol could function in a wireless environment. Networking is difficult enough in a predictable, managed *wired* environment. Talk about dropped packets! How can an IEEE 802 (Ethernet) variant function in free space? QoS refers, simply put, to reducing latency and jitter and avoiding dropping packets. This chapter alleviates those fears by addressing both legacy- and WiMAX-specific fixes to ensure carrier-grade performance in an otherwise hostile environment.

The Challenge

Mechanisms in the WiMAX MAC provide for differentiated QoS to support the different needs of different applications. For instance, voice and video require low latency but tolerate some error rate. In contrast, generic data applications cannot tolerate error, but latency is not critical. The standard accommodates voice, video, and other data transmissions by using appropriate features in the MAC layer; this is more efficient than using these features in layers of control overlaid on the MAC. In short, applying more bandwidth to the right channel at the right time reduces latency and improves QoS.

The WiMAX standard supports adaptive modulation, effectively balancing different data rates and link quality. The modulation method may be adjusted almost instantaneously for optimum data transfer. WiMAX is able to dynamically shift modulations from 64-QAM to QPSK via 16-QAM, displaying its ability to overcome QoS issues with dynamic bandwidth allocation over the distance between the BS and the SS.

Adaptive modulation allows efficient use of bandwidth and a broader customer base. The standard also supports both FDD and TDD. FDD, the legacy duplexing method, has been widely deployed in cellular telephony. It requires two channel pairs, one for transmission and one for reception, with some frequency separation between them to mitigate self-interference. A TDD system can dynamically allocate upstream and downstream bandwidth, depending on traffic requirements.[1]

Legacy QoS Mechanisms

The following paragraphs describe legacy mechanisms.

FDD/TDD/OFDM

WiMAX incorporates a number of time-proven mechanisms to ensure good QoS. Most notable are TDD, FDD, FEC, FFT, and OFDM. The WiMAX standard provides flexibility in spectrum usage by supporting both FDD and TDD. Thus, it can operate in both FDD/OFDM and TDD/OFDM modes. It supports two types of FDD: continuous FDD and burst FDD.

In continuous FDD, the upstream and downstream channels are located on separate frequencies, and all CPE stations can transmit and receive simultaneously. The downstream channel is always on, and all stations are always listening to it. Traffic is sent on this channel in a broadcast manner using TDM. The upstream channel is shared using TDMA, and the BS is responsible for allocating bandwidth to the stations.

In burst FDD, the upstream and downstream channels are located on separate frequencies. In contrast to continuous FDD, not all stations can transmit and receive simultaneously. Those that can transmit and receive simultaneously are referred to as full-duplex capable stations while those that cannot are referred to as half-duplex capable stations.

A TDD frame has a fixed duration and contains one downstream subframe and one upstream subframe. The two subframes are separated by a guard time called transition gap (TG), and the bandwidth that is allocated to each subframe is adaptive. The TDD subframe is illustrated in Figure 5-1.

---

[1]Dean Chang, "IEEE 802.16 Technical Backgrounder," Rev. 3, *IEEE* (2002): 3.

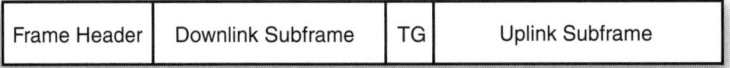

| Frame Header | Downlink Subframe | TG | Uplink Subframe |

Within a TDD downlink subframe, transmissions coming from the BS are organized into different modulation and FEC groups. The subframe header, called the FCH, consists of a preamble field, a PHY control field, and a MAC control field. The PHY control field is used for physical information, such as the slot boundaries, destined for all stations. It contains a map that defines where the physical slots for the different modulation/FEC groups begin.

The groups are listed in ascending modulation order, with QPSK first, followed by 16-QAM and then 64-QAM. Each CPE station receives the entire DL frame, decodes the subframe, and looks for MAC headers indicating data for the station. The DL data is always FEC coded. Payload data is encrypted, but message headers are unencrypted. The MAC control is used for MAC messages destined for multiple stations.

This variation uses burst single-carrier modulation with adaptive burst profiling in which transmission parameters, including the modulation and coding schemes, may be adjusted individually to each SS on a frame-by-frame basis. Channel bandwidths of 20 or 25 MHz (typical United States allocation) or 28 MHz (typical European allocation) are specified. Randomization is performed for spectral shaping and to ensure bit transitions for clock recovery.[2]

Forward Error Correction (FEC)

WiMAX utilizes FEC, a technique that doesn't require the transmitter to retransmit any information that a receiver uses for correcting errors incurred in transmission over a communication channel. The transmitter usually uses a common algorithm and embeds sufficient redundant information in the data block to allow the receiver to cor-

[2]Roger Marks, Carl Eklund, Kenneth Stanwood, and Stanley Wang, "IEEE 802.16: A Technical Overview of the WirelessMAN Air Interface for Broadband Wireless Access," *IEEE Communications*, June 2002, 98–107.

rect. Without FEC, error correction would require the retransmission of whole blocks or frames of data, resulting in added latency and a subsequent decline in QoS.[3]

Need QoS? Throw more bandwidth at it!

Throughput and latency are two essentials for network performance. Taken together, these elements define the "speed" of a network. Whereas throughput is the quantity of data that can pass from source to destination in a specific time, round-trip latency is the time it takes for a single data transaction to occur (the time between requesting data and receiving it). Latency can also be thought of as the time it takes from data send-off on one end to data retrieval on the other (from one user to the other). Therefore, the better throughput (bandwidth) management, the better the QoS.[4]

Bandwidth Is the Answer—What Was the Question?

To ensure consistent QoS, WiMAX's unique approach is to ensure consistent bandwidth. How is that achieved?

Bandwidth Requests and Grants   The WiMAX MAC accommodates two classes of SS that are differentiated by their ability to accept bandwidth grants for a single connection or for the SS as a whole. Both classes of SS request bandwidth per connection to allow the BS UL scheduling algorithm to properly consider QoS when allocating bandwidth. The two classes are GPC, where the BS grants bandwidth explicitly to each station, and GPSS, where bandwidth is granted to all connections belonging to the station.

The two classes of SS allow a trade-off between simplicity and efficiency. The need to explicitly grant extra bandwidth for RLC and requests, coupled with the likelihood of more than one entry per SS, makes GPC less efficient and scalable than GPSS. Additionally, the ability of the GPSS SS to react more quickly to the needs of the PHY

---

[3]Ibid., 118–119.

[4]"Low Latency—The Forgotten Piece of the Mobile Broadband Puzzle," white paper from Flarion, February 2003, www.flarion.com.

Table 5-1

Grants and
Requests for
Bandwidth to
Maintain Good
QoS

Class	Description
GPC	Bandwidth is granted explicitly to a connection, and the SS uses the grant only for that connection. RLC and other management protocols use bandwidth explicitly allocated to the management connections.
GPSS	SSs are granted bandwidth aggregated into a single grant to the SS itself. The GPSS SS needs to be more intelligent in its handling of QoS. It will typically, but need not, use the bandwidth for the connection that requested it. For instance, if the QoS situation at the SS has changed since the last request, the SS has the option of sending the higher QoS data along with a request to replace this bandwidth stolen from a lower QoS connection. The SS could also use some of the bandwidth to react more quickly to changing environmental conditions by sending, for instance, a DBPC-REQ message.

and those of connections enhances system performance. GPSS is the only class of SS allowed with the 10–66 GHz PHY. This is detailed in Table 5-1.

With both classes of grants, the WiMAX MAC uses a self-correcting protocol rather than an acknowledged protocol. This method uses less bandwidth. Furthermore, acknowledged protocols can take additional time, potentially adding delay. The bandwidth requested by an SS for a connection may not be available for a number of reasons:

- The BS did not see the request due to irrecoverable PHY errors or collision of a contention-based reservation.
- The SS did not see the grant due to irrecoverable PHY errors.
- The BS did not have sufficient bandwidth available.
- The GPSS SS used the bandwidth for another purpose.

In the self-correcting protocol, all of these anomalies are treated similarly. After a time-out appropriate for the QoS of the connection (or immediately, if the bandwidth was stolen by the SS for another purpose), the SS simply requests again. For efficiency, most bandwidth requests are incremental; that is, the SS asks for more bandwidth for a connection. However, for the self-correcting bandwidth request/grant mechanism to work correctly, the bandwidth requests must occasionally be aggregate; that is, the SS informs the BS of its total current bandwidth needs for a connection. This allows the BS to reset its perception of the SS's needs without a complicated protocol acknowledging the use of granted bandwidth.

The SS has many ways to request bandwidth, combining the determinism of unicast polling with the responsiveness of contention-based requests and the efficiency of unsolicited bandwidth. For continuous bandwidth demand, the SS need not request bandwidth; the BS grants it unsolicited. Bandwidth allocation and polling methods are detailed in Table 5-2.

To short-circuit the normal polling cycle, any SS with a connection running UGS can use the poll-me bit in the grant management

Table 5-2

*Bandwidth
Allocation
Polling
Methods*

Term	Description
Unicast polls	Used for inactive stations and active stations that have explicitly requested to be polled. If an inactive station does not require bandwidth allocation, it responds to the poll by returning a request for 0 bytes.
Multicast and broadcast polls	Used to poll a group of inactive stations when there is insufficient bandwidth to poll the stations individually. A CID identifies each active station, and certain CIDs are reserved for multicast and broadcast groups. When a multicast group is polled, the members of the group that require bandwidth allocation respond to the poll. They use the contention resolution algorithm to resolve any conflicts that arise from two or more stations transmitting at the same time. If a station does not need bandwidth allocation, it does nothing; it is not allowed to respond with a bandwidth allocation of zero, as with the case of the individual poll.
Station initiated polls	Used by stations to request that the BS poll them to request bandwidth allocation. Stations with active unsolicited grant service connections typically use the poll. A station initiating this type of poll sets a bit in the MAC header called the *poll-me* bit, typically to request to be polled more frequently in order to satisfy the QoS of the connection. When the base station receives the frame with the poll-me bit set, it polls the station individually.

Source: Ibe

subheader to inform the BS that it needs to be polled for bandwidth needs on another connection. The BS may choose to save bandwidth by polling SSs that have unsolicited grant services only when they have set the poll-me bit.

A more conventional way to request bandwidth is to send a bandwidth request MAC PDU that consists of simply the bandwidth request header and no payload. GPSS SSs can send this in any bandwidth allocation they receive. GPC terminals can send it in either a request interval or a data grant interval allocated to their basic connection. A closely related method of requesting data is to use a grant management subheader to piggyback a request for additional bandwidth for the same connection within a MAC PDU.[5] These types of services are detailed in Table 5-3.

UGS is tailored for carrying services that generate fixed units of data periodically. Here the BS schedules regularly, in a preemptive manner, grants of the size negotiated at connection setup without an explicit request from the SS. This eliminates the overhead and latency of bandwidth requests in order to meet the delay and delay jitter requirements of the underlying service. A practical limit on the delay jitter is set by the frame duration. If more stringent jitter requirements are to be met, output buffering is needed.

When used with UGS, the grant management subheader includes the poll-me bit as well as the slip indicator flag, which allows the SS to report that the transmission queue is backlogged due to factors such as lost grants or clock skew between the WiMAX system and the outside network.

The BS, upon detecting the slip indicator flag, can allocate some additional bandwidth to the SS, allowing it to recover the normal queue state. Connections configured with UGS are not allowed to utilize random access opportunities for requests.

The real-time polling service is designed to meet the needs of services that are dynamic in nature but offers periodic dedicated request opportunities to meet real-time requirements. Because the SS issues explicit requests, the protocol overhead and latency is

[5]Roger Marks, Carl Eklund, Kenneth Stanwood, and Stanley Wang, "IEEE 802.16: A Technical Overview of the WirelessMAN Air Interface for Broadband Wireless Access," 103–104.

Table 5-3

WiMAX Supports QoS Through Different Types of Service as Listed

Type of Service Supported by WiMAX	Description
UGS	Designed to support real-time service flows that generate fixed-size data packets, such as VoIP, on a periodic basis. Providing fixed-size data grants at periodic intervals eliminates the overhead and latency associated with requesting transmission channels.
Real-time polling service	Designed to support real-time service flows that generate variable-size data packets, such as MPEG video, on a periodic basis. The service period is defined to meet the flow's real-time needs and allow the station to specify the size of the desired grant.
UGS with activity detection	Designed to support UGS flows that may become inactive for a substantial length of time. This service is for stations that support real-time service when the flow is active and periodic unicast polls when the flow is inactive.
Non-real-time polling service	Designed to support non-real-time flows that require variable-size data grants, such as FTP, on a regular basis. The service offers unicast polls on a regular basis to ensure that flows receive request opportunity even during network congestion.
Best-effort service	Designed to provide efficient service to best-effort traffic.

Source: Ibe, Fixed Broadband Wireless Access Networks and Services

increased, but this capacity is granted only according to the real need of the connection. The real-time polling service is well suited for connections carrying services such as VoIP or streaming video or audio.

The non-real-time polling service is almost identical to the real-time polling service except that connections may utilize random access transmit opportunities for sending bandwidth requests. Typically, services carried on these connections tolerate longer delays and are rather insensitive to delay jitter. The non-real-time polling service is suitable for Internet access with a minimum guaranteed rate. A best-effort service has also been defined.

Neither throughput nor delay guarantees are provided. The SS sends requests for bandwidth in either random access slots or dedicated transmission opportunities. The occurrence of dedicated opportunities is subject to network load, and the SS cannot rely on their presence.

What Is FFT?   Electromagnetic waves have sines and cosines and are analog in nature while digital data is a stream of 1s and 0s resulting in square waves. How then can digital data be sent via an analog transmission? The theory is grounded on Fourier's Theorem (Emile Fourier was a French mathematician in the early 1800s), which proves that repeating, time-varying functions may be expressed as the sum of a possibly infinite series of sine and cosine waves. If 1,000 square waves are sent every second, the frequency components of sine waves are summed (1 KHz, 3 KHz, 5 KHz, and so on). Fast Fourier Transform is illustrated in Figure 5-2.

As the bit rate increases, the square wave frequency increases and the width of the square waves decreases. Eventually, narrower

Figure 5-2
Fast Fourier
Transform (FFT)

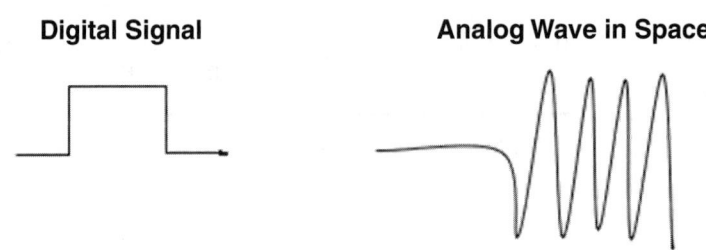

Digital Signal

Analog Wave in Space

Fast Fourier Transform

square waves require sine waves of even higher frequency to form the digital signal (read N^2). FFT makes these computations more efficient by reducing the computation to NlogN. Very simply put, FFT makes the transmission of digital data (square waves) over the airwaves more efficient.[6]

QPSK Versus QAM

Rather than attempting to be all things to all subscribers, WiMAX delivers a gradation of QoS dependent on distance of the SS from the BS: The greater the distance, the lower the guarantee of QoS. WiMAX utilizes three mechanisms for QoS; from highest to lowest, these mechanisms are 64-QAM, 16-QAM, and QPSK. Figure 5-3 illustrates modulation schemes.

By using a robust modulation scheme, WiMAX delivers high throughput at long ranges with a high level of spectral efficiency that is also tolerant of signal reflections. Dynamic adaptive modulation allows the BS to trade throughput for range. For example, if the BS cannot establish a robust link to a distant subscriber using the highest order modulation scheme, 64-QAM, the modulation order is

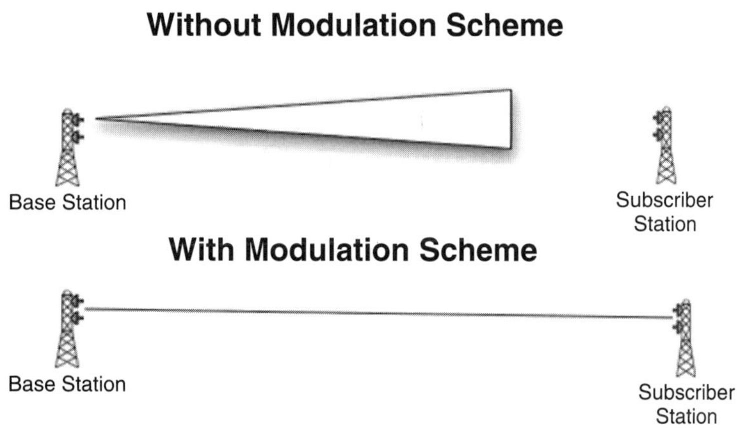

Figure 5-3
Modulation schemes focus the signal over distance.

Without Modulation Scheme

Base Station   Subscriber Station

With Modulation Scheme

Base Station   Subscriber Station

---

[6]Randall Nichols and Panos Lekkas, *Wireless Security: Models, Threats and Solutions* (New York: McGraw-Hill, 2002), 283.

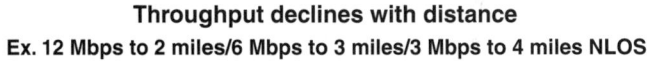

Figure 5-4
Modulation
schemes ensure
a quality signal
is delivered over
distance by
decreasing
throughput.

Throughput declines with distance
Ex. 12 Mbps to 2 miles/6 Mbps to 3 miles/3 Mbps to 4 miles NLOS

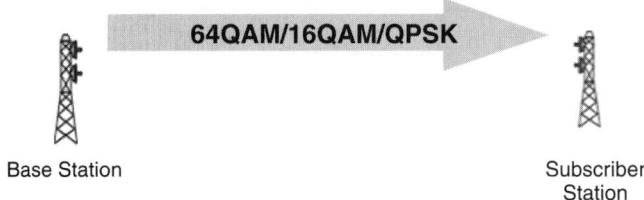

reduced to 16-QAM or QPSK, which reduces throughput and increases effective range. Figure 5-4 demonstrates how modulation schemes ensure throughput over distance.

QPSK and QAM are the two leading modulation schemes for WiMAX. In general the greater the number of bits transmitted per symbol, the higher the data rate is for a given bandwidth. Thus, when very high data rates are required for a given bandwidth, higher-order QAM systems, such as 16-QAM and 64-QAM, are used. 64-QAM can support up to 28 Mbps peak data transfer rates over a single 6 MHz channel. However, the higher the number of bits per symbol, the more susceptible the scheme is to intersymbol interference (ISI) and noise. Generally the signal-to-noise ratio (SNR) requirements of an environment determine the modulation method to be used in the environment. QPSK is more tolerant of interference than either 16-QAM or 64-QAM. For this reason, where signals are expected to be resistant to noise and other impairments over long transmission distances, QPSK is the normal choice.[7]

Multiplexing in OFDM

As shown in Figure 5-5, an efficient OFDM implementation converts a serial symbol stream of QPSK or QAM data into a size M parallel stream. These M streams are then modulated onto M subcarriers via the use of size N $(N{\leq}M)$ inverse FFT. The N outputs of the inverse FFT are then serialized to form a data stream that can then be mod-

[7]Oliver C. Ibe, *Fixed Broadband Wireless Access Networks and Services* (New York: John Wiley & Sons, 2002), 118–119.

ulated by a single carrier. Note that the N-point inverse FFT could modulate up to N subcarriers. When M is less than N, the remaining $N - M$ subcarriers are not in the output stream. Essentially, these have been modulated with amplitude of zero.

Although it would seem that combining the inverse FFT outputs at the transmitter would create interference between subcarriers, the orthogonal spacing allows the receiver to perfectly separate out each subcarrier. Figure 5-6 illustrates the process at the receiver. The received data is split into N parallel streams that are processed with

Figure 5-5
Block diagram
of a simple
OFDM
transmitter

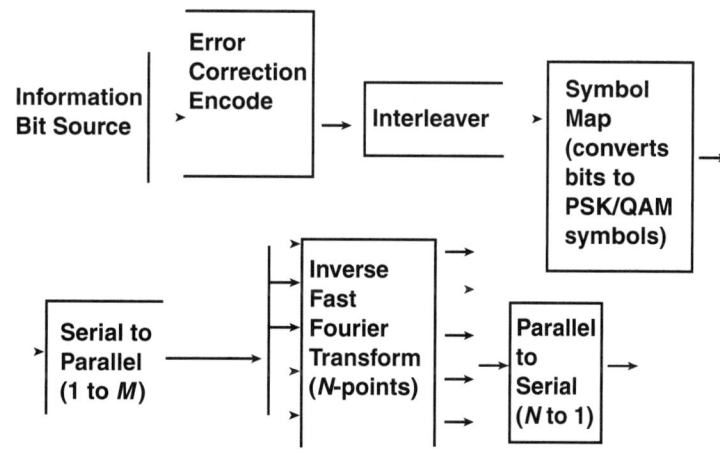

Figure 5-6
Block diagram
of a simple
OFDM receiver

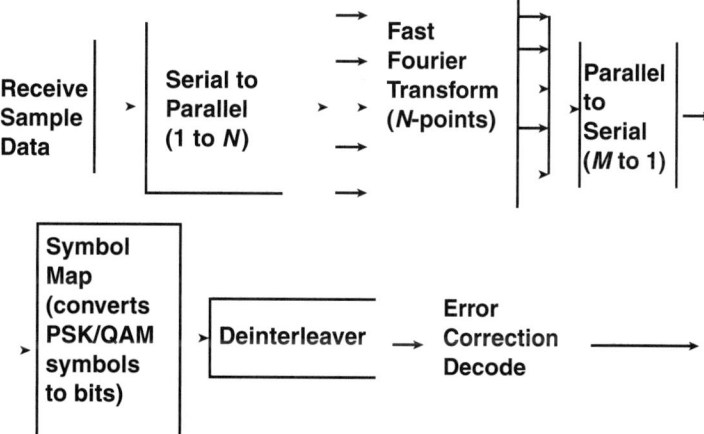

a size N FFT. The size N FFT efficiently implements a bank of filters, each matched to N possible subcarriers. The FFT output is then serialized into a single stream of data for decoding. Note that when M is less than N, in other words fewer than N subcarriers are used at the transmitter, the receiver only serializes the M subcarriers with data.

What OFDM Means to WiMAX

To the telecommunications industry, the "big so what?!" of WiMAX is that an WiMAX OFDM-based system can squeeze a 72 Mbps uncoded data rate (~100 Mbps coded) out of 20 MHz of channel spectrum. This translates into a spectrum efficiency of 3.6 bps per Hz. If five of these 20 MHz channels are contained within the 5.725 to 5.825 GHz band, giving a total band capacity of 360 Mbps (all channels added together with 1× frequency reuse). With channel reuse and through sectorization, the total capacity from one BS site could potentially exceed 1 Gbps.[8]

OFDM has manifold advantages in WiMAX, but among the more notable advantages is greater spectral efficiency. This is especially important in licensed spectrum use, where bandwidth and spectrum can be expensive. Here, OFDM delivers more data per spectrum dollar. In unlicensed spectrum applications, OFDM mitigates interference from other broadcasters due to its tighter beam width (less than 28 Mhz) and guardbands, as well as its dispersal of the data across different frequencies so that if one flow is "stepped on" by an interfering signal, the rest of the data is delivered on other frequencies.

QoS: Error Correction and Interleaving

Error correcting coding builds redundancy into the transmitted data stream. This redundancy allows bits that are in error or even missing to be corrected. The simplest example would be to simply repeat

---

[8]Kevin F.R. Suitor, "The Road to Broadband Wireless," white paper from Redline Communications, July, 2004, www.redlinecommunications.com.

the information bits. This is known as a repetition code. Although the repetition code is simple in structure, more sophisticated forms of redundancy are typically used because they can achieve a higher level of error correction. For OFDM, error correction coding means that a portion of each information bit is carried on a number of subcarriers; thus, if any of these subcarriers has been weakened, the information bit can still arrive intact.

Interleaving is the other mechanism used in OFDM systems to combat the increased error rate on the weakened subcarriers. Interleaving is a deterministic process that changes the order of transmitted bits. For OFDM systems, this means that bits that were adjacent in time are transmitted on subcarriers that are spaced out in frequency. Thus errors generated on weakened subcarriers are spread out in time; that is, a few long bursts of errors are converted into many short bursts. Error correcting codes then correct the resulting short bursts of errors.

QoS Measures Specific to the WiMAX Specification

WiMAX employs both legacy and next generation QoS measures. The following sections will focus on next generation QoS measures peculiar to WiMAX.

Theory of Operation

WiMAX QoS mechanisms function in both UL and DL frames through the SS and the BS. The WiMAX specification for QoS include the following:

- A configuration and registration function for preconfiguring SS-based QoS service flows and traffic parameters

- A signaling function for dynamically establishing QoS-enabled service flows and traffic parameters

- Utilization of MAC scheduling and QoS frame parameters for UL service flows

- Utilization of QoS traffic parameters for DL service flows

- Grouping of service flow properties into named service classes, so upper-Ayer entities and external applications (at both the SS and BS) may request service flows with desired QoS parameters in a globally consistent way

The principal mechanism for providing QoS is to associate packets traversing the MAC interface into a service flow as identified by the CID. A service flow is a unidirectional flow of packets that is provided a particular QoS (see Chapter 4). The SS and BS provide this QoS, according to the QoS parameter set defined for the service flow.

The primary purpose of the QoS features defined here is to define transmission ordering and scheduling on the air interface. However, these features often need to work in conjunction with mechanisms beyond the air interface in order to provide end-to-end QoS or to police the behavior of SSs.

Service flows in both the UL and DL direction may exist without actually being activated to carry traffic. All service flows have a 32-bit service flow ID (SFID); admitted and active flows also have a 16-bit CID.

Service Flows

A service flow is a MAC transport service that provides unidirectional transport of packets either to UL packets transmitted by the SS or to DL packets transmitted by the BS. A service flow is characterized by a set of QoS Parameters, such as latency, jitter, and throughput assurances. In order to standardize operation between the SS and BS, these attributes include details of how the SS requests UL bandwidth allocations and how the BS UL scheduler is expected to behave. The different elements of service flows employed by WiMAX are defined in Table 5-4. The three types of service flows are listed in Table 5-5.

Table 5-4

Elements of
the Service
Flow

Element	Description
SFID	The principal identifier for the service flow in the network. A service flow has at least an SFID and an associated direction.
CID	The mapping to an SFID that exists only when the connection has an admitted or active service flow.
ProvisionedQoSParamSet	A QoS parameter set provisioned via means outside of the scope of the standard, such as the network management system.
AdmittedQoSParamSet	A set of QoS parameters for which the BS (and possibly the SS) is reserving resources. The principal resource to be reserved is bandwidth. This set also includes an additional memory or time-based resource required to subsequently activate the flow.
ActiveQoSParamSet	A set of QoS parameters defining the service actually being provided to the service flow. Only an active service flow may forward packets.
Authorization Module	A logical function within the BS that approves or denies every change to QoS Parameters and Classifiers associated with a service flow. As such, it defines an "envelope" that limits the possible values of the AdmittedQoSParamSet and ActiveQoSParamSet.

Table 5-5

Types of
Service Flows

Service Flow	Description
Provisioned	This service flow is known via provisioning by, for example, the network management system. Its AdmittedQoSParamSet and ActiveQoSParamSet are both null.
Admitted	This service flow has resources reserved by the BS for its AdmittedQoSParamSet, but these parameters are not active. Some other mechanism has provisioned or may have signaled admitted service flows.
Active	This service flow has resources committed by the BS for its ActiveQoSParamSet. Its ActiveQoSParamSet is non-null.

The Object Model

The major objects of the architecture are represented by named rectangles, as illustrated in Figure 5-7. Each object has a number of attributes; the attribute names that uniquely identify the object are underlined. Optional attributes are denoted with brackets. The relationship between the number of objects is marked at each end of the associated line between the objects. For example, a service flow may be associated with from 0 to N (many) PDUs, but a PDU is associated with exactly one service flow. The service flow is the central concept of the MAC protocol. It is uniquely identified by a 32-big SFID. Service flows may be in either the UL or DL direction. Admitted and active service flows are mapped to a 16-bit CID.

A CS process submits outgoing user data to the MAC SAP for transmission on the MAC interface. The information delivered to the MAC SAP includes the CID identifying the connection across which the information is delivered. The service flow for the connection is mapped to MAC connection identified by the CID.

The service class is an optional object that may be implemented at the BS. It is referenced by an ASCII name, which is intended for pro-

Figure 5-7
Theory of
operation
object model
(Source: IEEE)

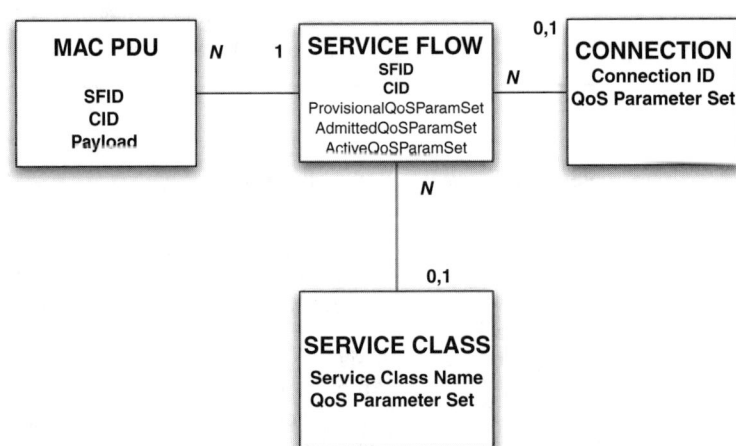

visioning purposes. A service class is defined in the BS to have a particular QoS parameter set. The QoS parameter sets of a service flow may contain a reference to the service class name as a *macro* that selects all of the QoS parameters of the service class. The service flow QoS parameter sets may augment and even override the QoS parameter settings of the service class, subject to authorization by the BS.

Service Classes

The service class performs two functions. First, it allows operators to shift configuring service flows from the provisioning server to the BS. Operators provision the SSs with the service class name; full implementation of the name is configured at the BS. This allows operators to modify the implementation of a given service to local circumstances without changing SS provisioning.

Second, it allows higher-layer protocols to create a service flow by its service class name. For example, telephony signaling may direct the SS to instantiate any available provisioned service flow of class "G.711."

Any service flow may have its QoS parameter set specified in any of three ways: first, by explicitly including all traffic parameters; second, by indirectly referring to a set of traffic parameters by specifying a service class name; and third, by specifying a service class name along with modifying parameters.

Authorization

An Authorization Module will approve every change to the service flow QoS parameters. This includes every DSA-REQ message to change a QoS parameter set of an existing service flow. Such changes create a new service flow, and every DSC-REQ message changes a QoS parameter set of an existing service flow. Such changes include requesting an admission control decision (for example, setting the AdmittedQoSParamSet) and requesting activation of a service flow (for example, setting the ActiveQoSParamSet). The Authorization Module also checks reduction requests regarding the resources to be admitted or activated. This is further defined in Table 5-6.

Prior to initial connection setup, the BS retrieves the Provisional QoS parameter set for an SS that is handed to the Authorization Module in the BS. The BS will be capable of caching the Provisional QoS parameter set and will be able to use this information to authorize dynamic flows that are a subset of the Provisional QoS parameter set.

Types of Service Flows

The three types of service flows are described in Table 5-7.

Service Flow Creation   During provisioning, a service flow is instantiated and gets a service flow ID (SFID) and a provisioned

Table 5-6

WiMax QoS
Authorization
Models

Type	Description
Static authorization	Stores provisioned status of all "deferred" service flows. Admission and activation requests for these provisioned service flows shall be permitted as long as the Admitted QoS parameter set is a subset of the Provisioned QoS parameter set, and the Active QoS parameter set is a subset of the Admitted QoS parameter set. Requests to change the Provisioned QoS parameter set will be refused, as will requests to create new dynamic service flows. Static authorization defines a static system where all possible services are defined in the initial configuration of each SS.
Dynamic authorization	Communicates through a separate interface to an independent policy server that provides authorization module with advance notice of upcoming admission and activation requests and specifies proper authorization action to be taken on requests. The Authorization Module then checks admission and activation requests from an SS to ensure the ActiveQoSParamSet being requested is a subset of the set provided by the policy server. Admission and activation requests from an SS that are signaled in advance by the external policy server are permitted.

type. Enabling service flows follows the transfer of the operational parameters.

Service Flow Creation—SS-Initiated Either the BS or the SS may initiate the service flows. A DSA-REG from an SS (see Figure 5-8) contains a service flow reference and QoS parameter set

Table 5-7

Types of
Service Flows

Service Flow	Description
Provisional service flows	A service flow that is provisioned but not immediately activated. The network assigns an SFID to provisional service flows.
Admitted service flows	A two-phase activation model. First, the resources for a call are admitted; once the end-to-end negotiation is completed, the resources are activated.
Active service flow	A service flow that has a non-null Active-QoSParamSet.

Figure 5-8
DSA message
flow—SS-initiated
(Source: IEEE)

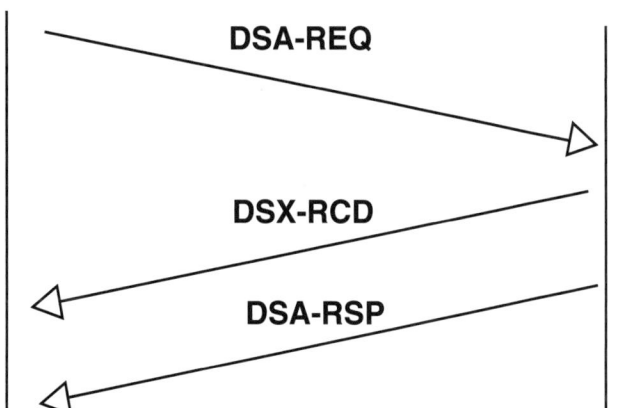

(marked either for admission-only or for admission and activation). The BS responds with a DSA-RSP indicating acceptance or rejection.

Dynamic Service Flow Creation—BS-Initiated A DSA-REQ from a BS (see Figure 5-9) contains an SFID for one UL or one DL service flow, possibly its associated CID, and a set of active or admitted QoS parameters. The SS responds with DSA-RSP indicating acceptance or rejection.

Service Flow Management

Service flows may be created, changed, or deleted. This is accomplished through a series of MAC management messages listed in Table 5-8.

As Figure 5-10 illustrates, the null state implies no service flow exists that matches the SFID in a message. Once the service flow

Figure 5-9
DSA message
flow—
BS-initiated
(Source: IEEE)

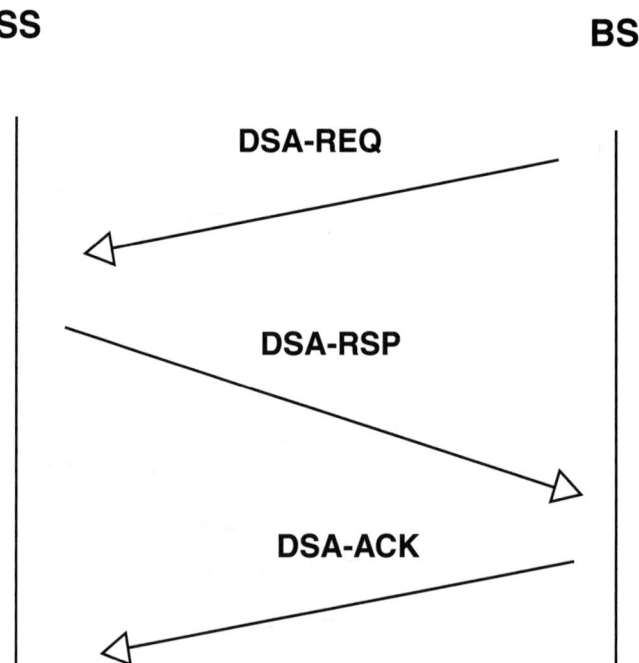

SS   BS

DSA-REQ

DSA-RSP

DSA-ACK

Table 5-8

Service Flow
Messages

Type Message	Description
Dynamic Service Change (DSC)	Changes existing service flow
Dynamic Service Delete (DSD)	Deletes existing service flows
Dynamic Service Activate (DSA)	Activates a service flow

exists, it is operational and has an assigned SFID. In steady-state
operation, a service flow resides in a nominal state.[9]

Conclusion

Service providers considering a WiMAX solution should take comfort
in the many measures, both legacy- and WiMAX-specific, that focus
on QoS issues. As the transmission is over free space, it is important
that the QoS measures account for what is perhaps the most difficult
of datacom environments. Legacy measures (such as TDD, FDD, and
OFDM) uniquely address QoS issues for this protocol. Object models
and dynamic service flows along with QoS parameters ensure good
QoS over the airwaves using WiMAX.

Figure 5-10
Dynamic service
flow overview
(Source: IEEE)

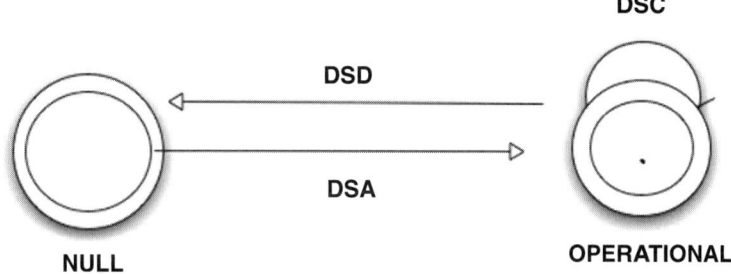

---

[9]"802.16-2004 IEEE Standard for Local and Metropolitan Area Networks, Part 16, Air
Interface for Fixed Broadband Wireless Access Systems," June 24, 2004, 219–217.

Dealing with Interference with WiMAX

Interference—Some Assumptions

The primary objection to wireless systems is the concern that there are or will soon be too many operators on the same frequency, which will cause so much interference that the technology will become unusable. This issue is not that simple.

Such an assumption relies largely on the use of unlicensed spectrum, where, according to Larry Lessig's "tragedy of the commons" scenario,[1] multiple operators broadcast on the same unlicensed (read "free") spectrum, ultimately rendering it useless. Although this scenario may already be evident in the case of Wi-Fi variants (largely limited to the 2.4 GHz range), WiMAX is considerably different. WiMAX currently has no problems, only solutions.

Since 1927, interference protection has always been at the core of federal regulators' spectrum mission. The Radio Act of 1927 empowered the Federal Radio Commission to address interference concerns. This act primarily focused on three parameters: location, frequency, and power. The technology of the time did not permit consideration of a fourth element: time. In the modern sense, one might consider that a spectrum used by cell phones in a metropolitan area (dense population with millions of users) would command a very high price at a spectrum auction. At the other end of the "spectrum," a frequency band, say 2.5 GHz, in an exurban or rural market may go for very little money at an auction or at resell by a spectrum broker. It is entirely possible that the wireless service provider may find a very low cost licensed spectrum and enjoy a protected spectrum, which will largely negate the concern over interference from other broadcasters (the purpose of the Radio Act of 1927 in the first place).

Defining Interference or "Think Receiver"

The Interference Protection Working Group of the FCC's Spectrum Policy Task Force recommends that the FCC should consider using

---

[1]Larry Lessig, *The Future of Ideas* (San Francisco: Vintage Press), October 2002.

the "interference temperature" metric to quantify and manage interference. "Interference temperature" is a measure of radio frequency (RF) power (power generated by other emitters and noise sources) available at a *receiving* antenna to be delivered to a receiver. More specifically, it is the temperature equivalent of the RF power available at a receiving antenna per unit bandwidth, measured in units of degrees Kelvin. As conceptualized by the FCC, the terms "interference temperature" and "antenna temperature" are synonymous. The term "interference temperature" is more descriptive for interference management.

Interference temperature can be calculated as the power received by an antenna (watts) divided by the associated RF bandwidth (hertz) and a term known as Boltzman's Constant (equal to 1.3807 wattsec per °Kelvin). Alternatively, it can be calculated as the power flux density available at a receiving antenna (watts per meter squared), multiplied by the effective capture area of the antenna (meter squared), with this quantity divided by the associated RF bandwidth (hertz) and Boltzman's Constant. An "interference temperature density" can also be defined as the interference temperature per unit area, expressed in units of °Kelvin per meter squared and calculated as the interference temperature divided by the effective capture area of the receiving antenna (determined by the antenna gain and the received frequency). Interference temperature density can be measured for particular frequencies using a reference antenna with known gain. Thereafter, it can be treated as a signal propagation variable independent of receiving antenna characteristics.

As illustrated in Figure 6-1, interference temperature measurements can be taken at receiver locations throughout the service areas of protected communications systems, thus estimating the real-time conditions of the RF environment.[2]

Forms of Interference

Interference can be classified into two broad categories: co-channel (CoCh) interference (internal) and out-of-channel interference

---

[2]Michael Powell, "Broadband Migration—New Directions in Wireless Policy" (speech to Silicon Flatirons Conference, University of Colorado, Boulder, October 30, 2002).

Figure 6-1
Interference
temperature
illustrated
(Source: FCC)

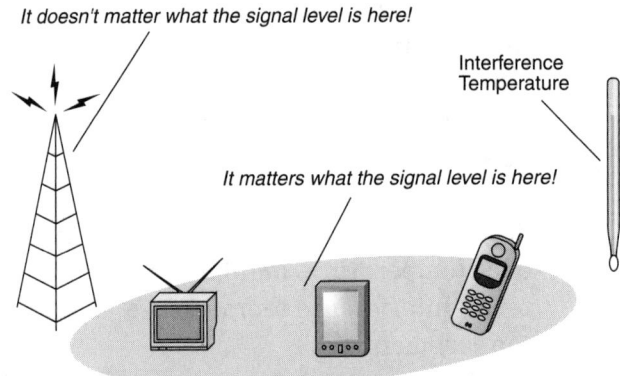

(external). These forms of interference manifest themselves as shown in Figure 6-2.

Figure 6-2 illustrates a simplified example of the power spectrum of the desired signal and CoCh interference. Note that the channel bandwidth of the CoCh interferer may be wider or narrower than the

Figure 6-2
Forms of
interference
(Source: IEEE)

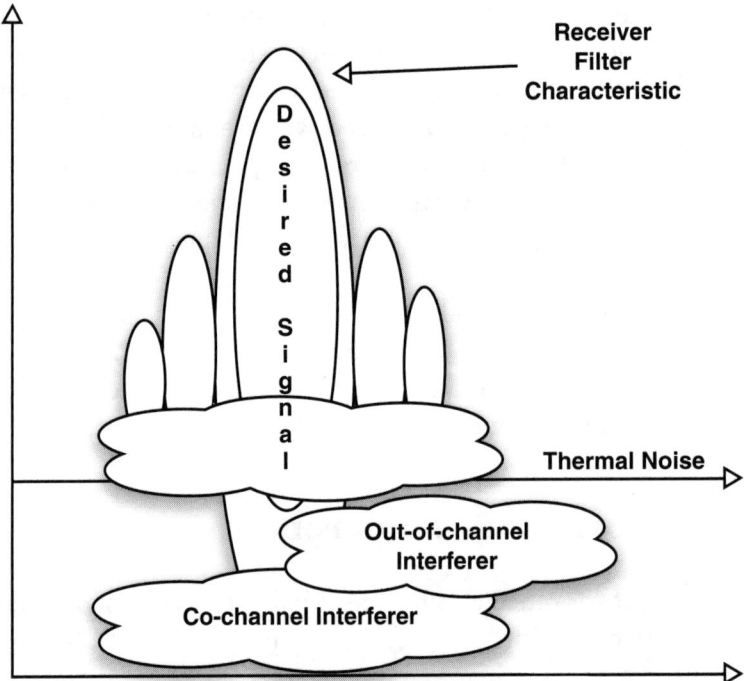

desired signal. In the case of a wider CoCh interferer (as shown), only a portion of its power will fall within the receiver filter bandwidth. In this case, the interference can be estimated by calculating the power arriving at the receive (Rx) antenna and then multiplying by a factor equal to the ratio of the filter's bandwidth to the interferer's bandwidth.

An out-of-channel interferer is also shown. Here, two sets of parameters determine the total level of interference. First, a portion of the interferer's spectral sidelobes or transmitter output noise floor falls CoCh to the desired signal, that is, within the receiver filter's passband. This can be treated as CoCh interference. It cannot be removed at the receiver; its level is determined at the interfering transmitter. By characterizing the power spectral density (psd) of sidelobes and output noise floor with respect to the main lobe of a signal, this form of interference can be approximately computed similarly to the CoCh interference calculation, with an additional attenuation factor due to the suppression of this spectral energy with respect to the main lobe of the interfering signal. Figure 6-3 details the relationship of these lobes to the transmitter.

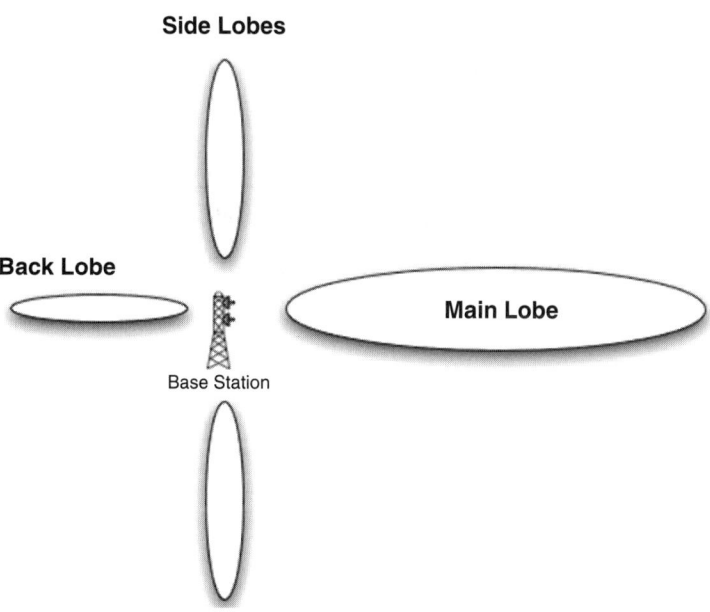

Figure 6-3
Main lobe, side lobes, and back lobe

Second, the receiver filter of the victim receiver does not completely suppress the main lobe of the interferer. No filter is ideal, and residual power passing through the stopband of the filter can be treated as additive to the CoCh interference present. The performance of the victim receiver in rejecting out-of-channel signals, sometimes referred to as *blocking performance*, determines the level of this form of interference. This form of interference can be simply estimated in a manner similar to the CoCh interference calculation, with an additional attenuation factor due to the relative rejection of the filter's stopband at the frequency of the interfering signal.

Cofrequency/Adjacent-Area Case   Operators are encouraged to arrive at mutually acceptable sharing agreements that would allow for the maximum provision of service by each licensee within its service area. Under the circumstances where a sharing agreement between operators does not exist or has not been concluded and where service areas are in close proximity, a coordination process should be employed.[3]

Countering Interference

Four parameters are brought under the control of network planners to minimize external sources of interference:

- Channel/band/frequency
- Distance to the interference (farther is better)/distance to intended signal (closer is better)
- Power levels (lower is better)
- Antenna technology

---

[3]IEEE 802.16.2-2004, "Coexistence of Fixed Broadband Wireless Access Systems," March 17, 2004, 77–78.

Changing Channels Within the ISM or U-NII Bands

WiMAX's specification calls for, depending on the variant, a frequency spread from 2–66 GHz (contrast with Wi-Fi's, limited to 2.4 GHz). Given that frequency spread, a for-profit service provider would be wise to consider a low-cost licensed frequency and avoid altogether the discussion of interference from other service providers. The purpose of licensed frequency is to protect a broadcaster from other broadcasters interfering with his or her transmission. (This is the original intent of the Radio Act of 1927.)

Recent changes in FCC policy now dictate that spectrum holders may resell their unused spectrum to other broadcasters, thus opening that spectrum to other operators. The FCC even hints at forcing the resell of unused spectrum. See Chapter 10 for more information on the regulatory aspects of WiMAX.

The specifications for industrial, scientific, and medical (ISM) and unlicensed national information infrastructure (U-NII) stipulate multiple channels or frequencies. If interference is encountered on one frequency, the broadcaster can merely switch frequencies to a channel that is not being interfered with. ISM provides 11 overlapping channels (for North America): each channel is 22 MHz wide and is centered at 5 MHz intervals (beginning at 2.412 GHz and ending at 2.462 GHz). This means that only three channels (channels 1, 6, 11) do NOT overlap. Table 6-1 indicates the channels of the unlicensed ISM band.

802.11a provides 12 channels: each channel is 20 MHz wide and is centered at 20 MHz intervals (beginning at 5.180 GHz and ending at 5.320 GHz for the upper and middle U-NII bands, beginning at 5.745 GHz and ending at 5.805 GHz for the upper U-NII band). It is important to note that none of these channels overlap.[4]

---

[4]"A Comparison of 802.11a and 802.11b Wireless LAN Standards," white paper from Linksys, November 2004, www.linksys.com/products/images/wp_802.asp.

Table 6-1

Eleven
Channels of
the Unlicensed
ISM Band

Channel	Frequency (GHz)
1	2.412
2	2.417
3	2.422
4	2.427
5	2.432
6	2.437
7	2.442
8	2.447
9	2.452
10	2.457
11	2.462

Dealing with Distance

The delivery of an intelligible signal is a function of both the power of the signal and the distance between transmitter and receiver. A fundamental concept in any communications system is the link budget, a summation of all the gains and losses in a communications system. The link budget results in the transmit power required to present a signal with a given SNR at the receiver to achieve a target bit error rate (BER).

A signal on the same frequency as the WiMAX WMAN, for example, will not interfere if the source is too distant. That is, the interfering signal becomes too weak to present interference. In addition, if the distance between the BS and the subscriber device is greater than optimal, the signal weakens over the distance and becomes susceptible to interference, as the interfering signal is greater than the desired signal. Figure 6-4 illustrates coverage area using a series of cells.

Figure 6-4
Each cell
represents the
maximum range
of each BS.

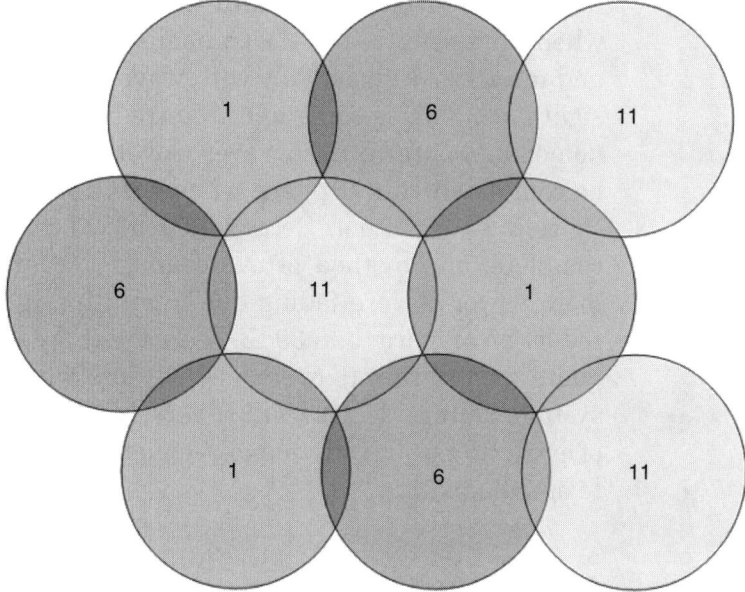

Engineering with Power   Power levels of the primary and interfering signals must also be taken into account. If the power level of the interfering signal gets close to the power level of the intended WiMAX signal, then interference will occur. The simplest solution is to increase the power level of the WiMAX signal in order to overcome the interfering signal. The limitation here is that the service provider must not interfere with licensed spectrum operators on similar (unlikely) spectrum.[5]

Internal (CoCH) Sources of Interference

Sometimes a wireless network's greatest interferer is itself. A number of challenges arise from within a wireless network due to the nature of wireless transmissions. These sources of interference include multipath interference and channel noise. Both can be engineered out of the network.

---

[5]See FCC Regulations, parts 15.247 and 15.407, www.fcc.gov.

Multipath Distortion and Fade Margin Multipath occurs when waves emitted by the transmitter travel along a different path and interfere destructively with waves traveling on a direct line-of-sight path. This is sometimes referred to as signal fading. This phenomenon occurs because waves traveling along different paths may be completely out of phase when they reach the antenna, thereby canceling each other. Because signal cancellation is almost never complete, one method of overcoming this problem is to transmit more power. Severe fading due to multipath can result in a signal reduction of more than 30dB. It is therefore essential to provide adequate link margin to overcome this loss when designing a wireless system. Failure to do so will adversely affect reliability. The amount of extra RF power radiated to overcome this phenomenon is referred to as fade margin.[6]

OFDM in Overcoming Interference

Very simply put, OFDM is a silver bullet used by WiMAX to overcome many forms of interference.

Multipath Challenges In an OFDM-based WMAN architecture, as well as in many other wireless systems, multipath distortion is a key challenge. This distortion occurs at a receiver when objects in the environment reflect a part of the transmitted signal energy. Figure 6-5 illustrates one such multipath scenario from a WMAN environment.

Multipath-reflected signals arrive at the receiver with different amplitudes, different phases, and different time delays. Depending on the relative phase change between reflected paths, individual frequency components will add constructively and destructively. Consequently, a filter representing the multipath channel shapes the frequency domain of the received signal. In other words, the receiver may see some frequencies in the transmitted signal that are attenuated and others that have a relative gain.

---

[6]Jim Zyren and Al Petrick, "Tutorial on Basic Link Budget Analysis," white paper from Intersil, June 1998, p. 2, www.intersil.com.

Figure 6-5
Multipath
reflections
(shown here)
create
intersymbol
interference (ISI)
in OFDM
receiver designs.

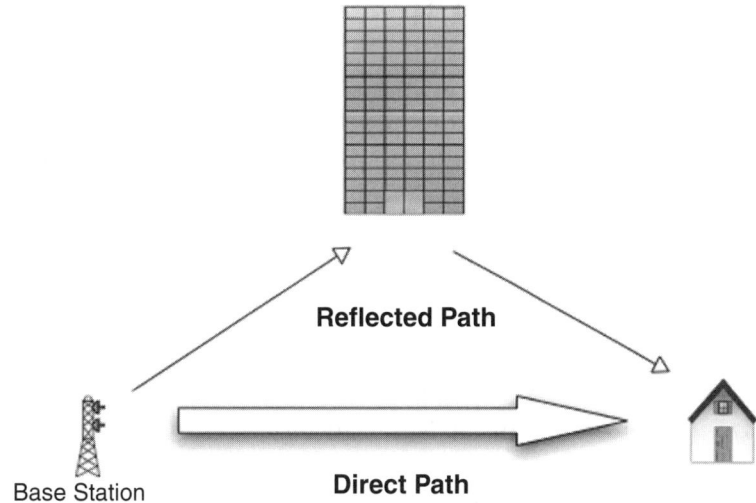

Reflected Path

Direct Path

Base Station

In the time domain, the receiver sees multiple copies of the signal with different time delays. The time difference between two paths often means that different symbols will overlap or smear into each other and create ISI. Thus, designers building WLAN architectures must deal with distortion in the demodulator.

OFDM relies on multiple narrowband subcarriers. In multipath environments, the subcarriers located at frequencies attenuated by multipath will be received with lower signal strength. The lower signal strength leads to an increased error rate for the bits transmitted on these weakened subcarriers.

Fortunately for most multipath environments, this affects only a small number of subcarriers and, therefore, only increases the error rate on a portion of the transmitted data stream. Furthermore, the robustness of OFDM in multipath can be dramatically improved with interleaving and error correction coding. Intersymbol interference is illustrated in Figure 6-6.

Figure 6-6
ISI

SHORT PATH

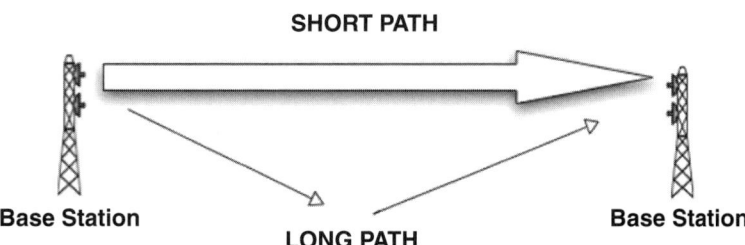

Base Station

LONG PATH

Base Station

Handling ISI

The time-domain counterpart of the multipath is ISI or smearing of one symbol into the next. OFDM handles this type of multipath distortion by adding a *guard interval* to each symbol. The guard interval is typically a cyclic or periodic extension of the basic OFDM symbol. In other words, it looks like the rest of the symbol but conveys no *new* information.

Because no new information is conveyed, the receiver can ignore the guard interval and still be able to separate and decode the subcarriers. When the guard interval is designed to be longer than any smearing due to the multipath channel, the receiver is able to eliminate ISI distortion by discarding the unneeded guard interval. Hence, ISI is removed with virtually no added receiver complexity.

It is important to note that discarding the guard interval does impact noise performance because the guard interval reduces the amount of energy available at the receiver for channel symbol decoding. In addition, it reduces the data rate, as no new information is contained in the added guard interval. Thus a good system design will make the guard interval as short as possible while maintaining sufficient multipath protection.

Why don't single carrier (SC not OFDM) systems also use a guard interval? Single carrier systems could remove ISI by adding a guard interval between each symbol. However, this has a much more severe impact on the data rate for single carrier systems than it does for OFDM. Because OFDM uses a bundle of narrowband subcarriers, it obtains high data rates with a relatively long symbol period because the frequency width of the subcarrier is inversely proportional to the symbol duration. Consequently, adding a short guard interval has little impact on the data rate.

Single carrier systems with bandwidths equivalent to OFDM must use much shorter duration symbols. Hence, adding a guard interval equal to the channel smearing has a much greater impact on data rate.[7]

---

[7]Steven Halford and Karen Halford, "OFDM Uncovered: The Architecture," white paper from CommsDesign, May 2, 2002, www.commsdesign.com/design_corner/OEG20020502S0013.

Mitigating Interference with Antenna Technology

New antenna technologies help reduce interference in WiMAX networks.

Multiple Antennas: AAS

One method of mitigating the effects of multipath is antenna diversity. Because the cancellation of radio waves is geometry dependent, using two (or more) antennas separated by at least half of a wavelength can drastically mitigate this problem. On acquisition of a signal, the receiver checks each antenna and simply selects the antenna with the best signal quality. This reduces but does not eliminate the required link margin that would otherwise be needed for a system that does not employ diversity.

The downside is this approach requires more antennas and a more complicated receiver design. Another method of dealing with the multipath problem is using an adaptive channel equalizer. Adaptive equalization can be used with or without antenna diversity. Figure 6-7 illustrates how adaptive antennas use beam forming to overcome interference.

WiMAX currently supports several multiple-antenna options including STC, MIMO antenna systems, and AAS. Table 6-2 illustrates the advantages of using multiple-antenna over single antenna technology.

A common scheme that exhibits both array gain and diversity gain is maximal ratio combining: this scheme combines multiple receive paths to maximize SNR. Selection diversity, on the other hand, primarily exhibits diversity gain. The signals are not combined; rather, the signal from the best antenna is chosen.

For AASs, multiple overlapped signals can be transmitted simultaneously using SDMA, a technique that exploits the spatial dimension to transmit multiple beams that are spatially separated. SDMA makes use of CCIR, diversity gain, and array gain.

Figure 6-7

Adaptive
antennas use
beam forming
to avoid
interference.

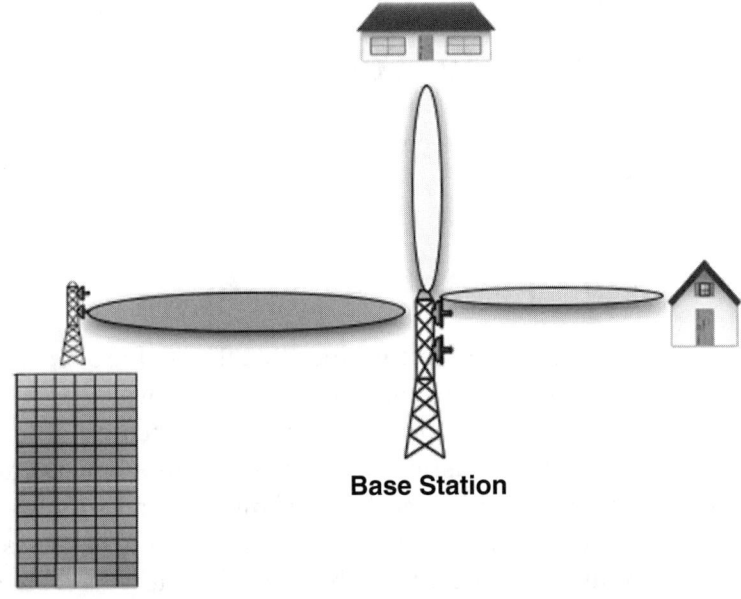

Base Station

Table 6-2

Advantages of
Using Multiple-
Antenna
Technology
Over Single
Antenna
Technology

Type Antenna	Description
Array gain	Gain achieved by using multiple antennas so that the signal adds coherently.
Diversity gain	Gain achieved by utilizing multiple paths so that a single bad path does not limit performance. Effectively, diversity gain refers to techniques at the transmitter or receiver to achieve multiple *looks* at the fading channel. These schemes improve performance by increasing the stability of the received signal strength in the presence of wireless signal fading. Diversity may be exploited in the spatial (antenna), temporal (time), or spectral (frequency) dimensions.
Co-channel Interference Rejection (CCIR)	Rejection of signals by using the different channel response of the interferers.

The higher performance and lower interference capabilities of MIMO and AAS make them attractive over other high-rate techniques for WiMAX systems in costly, licensed bands. A key advantage of transmit diversity is that it can be implemented at the BS, which can absorb higher costs of multiple antennas and associated RF chains. This shifts cost away from the SS, which enables faster market penetration of WiMAX products.[8]

Adaptive Antenna (AA) Techniques

AA directly affects coexistence because the RF energy radiated by transmitters is focused in specific areas of the cell, not radiated in all directions. Moreover, beam forming, with the goal of maximizing the link margin for any given user inside the cell coverage area at any given time, makes the AA beams' azimuth and elevation vary from time to time. Figure 6-8 explains interference vis-à-vis non-AAS cells.

Figure 6-8
Non-AAS cell

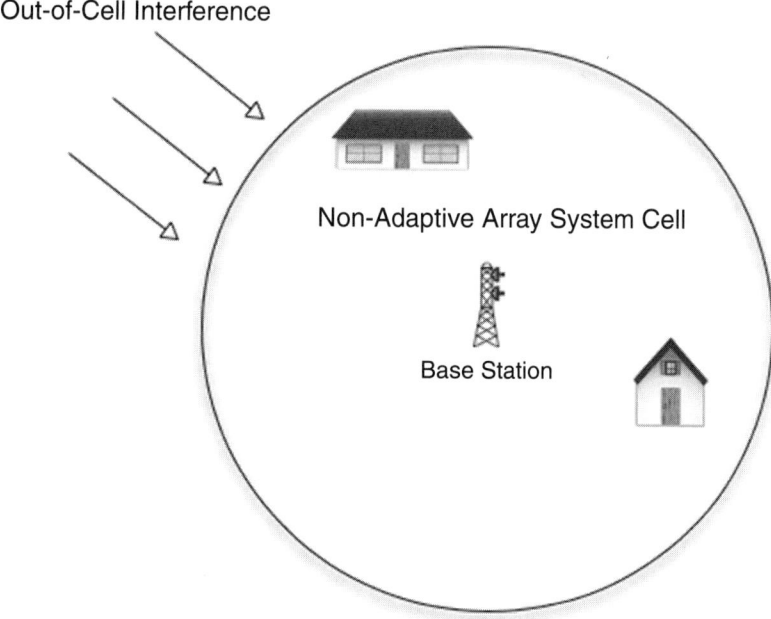

Out-of-Cell Interference

Non-Adaptive Array System Cell

Base Station

---

[8]Atul Salvekar, Sumeet Sandhu, Qinghua Li, Minh-Anh Vuong, and Xiashu Qian, "Multiple Antenna Technology in WiMAX Systems," *Intel Technology Journal* 8, no. 3 (August 20, 2004), http://developer.intel.com/technology/itj/2004/volume08issue03/art05_multiantenna/p01_abstract.htm.

This characteristic would play a major role in determining the likelihood of interference in both the adjacent area and adjacent frequency block coexistence scenarios. Although the worst-case alignment scenario may look prohibitive because beam forming may produce a higher gain in the wanted direction, the statistical factor introduced by using AA may allow an otherwise unacceptable coexistence environment to become tolerable. Figure 6-9 illustrates the advantages of AAS technology.

Other Characteristics of AAs   Other characteristics could supplement the improvement brought about by the statistical nature of AA operation and warrant further analysis.

Signal processing and the development of spatial signatures associated with the wanted stations may also help to discriminate against interferers in certain directions, further reducing the total impact of cumulative interference from neighboring systems in adja-

Figure 6-9
AAS cell: Note extended range and resistance to outside interference.

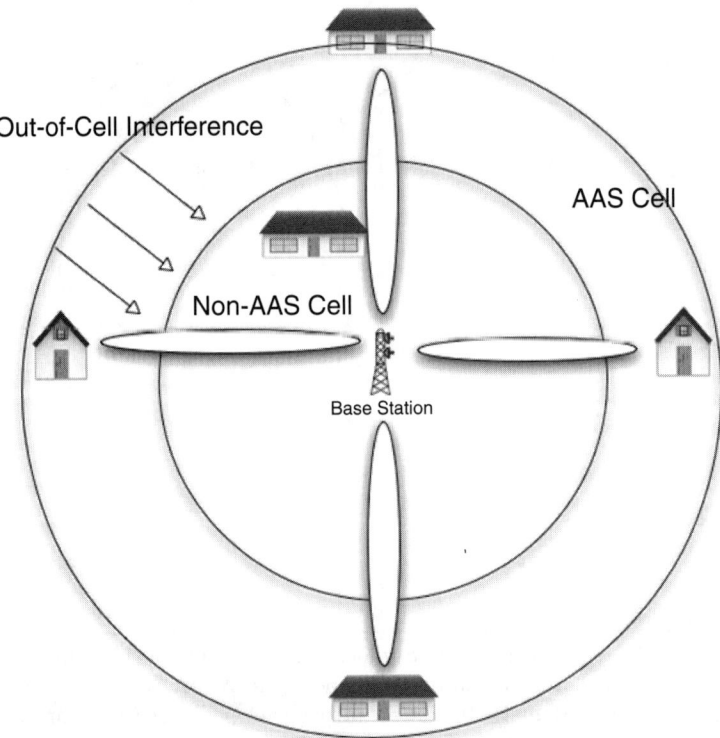

cent areas. For systems operating in adjacent frequencies, the loss of coherency in out-of-band operations reduces the AA gain toward the interferers/victims, which could reduce the amount of interference power.[9]

Dynamic Frequency Selection (DFS)

The WiMAX specification calls for a mechanism called DFS for use in unlicensed frequencies. This mechanism simply has the service flow shift to a different frequency if activity is detected on a primary frequency.

If You Want Interference, Call the Black Ravens

One of the author's first *real* jobs was intelligence officer, Tactical Electronic Warfare Squadron 135 (abbreviated VAQ-135 with the nickname "World Famous Black Ravens") of the United States Navy. This squadron flew the EA-6B tactical jamming aircraft. The airplane is equipped with ALQ-99 jamming system and is used tactically to jam enemy radar and radio communications. It has been rumored for many years that the squadron's four aircraft, strategically positioned, could shut down most of the electromagnetic spectrum of the United States (TV, radio, and so on). Figure 6-10 is a photo of the officers and men of VAQ-135 with the EA-6B in the background.

In a strategic role during the Cold War, the United States Air Force developed the B-52G, a bomber equipped with an extensive suite of electronic jamming equipment designed to defeat the Soviet

---

[9]"802.16.2™ IEEE Recommended Practice for Local and Metropolitan Area Networks Coexistence of Fixed Broadband Wireless Access Systems," *IEEE* (March 2004): 86–87.

Figure 6-10
EA-6B tactical jamming aircraft of the United States Navy, the "World Famous Black Ravens." Author Ohrtman is first on left, standing.

air defenses. This would require overwhelming air defense overlapping radar networks that operated at a variety of frequencies. It would also deliver overwhelming interference on air defense radio communications, making the airwaves unusable for the Soviets. By shutting down Soviet air defense radars and negating their ability to communicate by radio, the B-52G would clear a path for itself and other strategic bombers to targets for destruction by nuclear attack. A trivia question on student examinations at the United States Navy's Electronic Warfare School in the 1980s was "What is the electromagnetic coverage of the B-52G jamming system?" The correct answer was "DC (Direct Current) to Daylight."

Security and 802.16 WiMAX

Security in WiMAX Networks

A major objection service providers have toward broadband wireless access networks is security. Will the wireless protocol provide adequate security to prevent theft of service, thus protecting their investment in the wireless infrastructure? Will the privacy of their subscribers be protected from hackers who might ultimately perpetrate identity theft? The WiMAX specification offers some very powerful security measures, making casual theft of service impossible. WiMAX subscribers need not fear for their privacy while utilizing this wireless service.

The Security Sublayer

The WiMAX specification includes a security sublayer that provides subscribers with privacy across the fixed broadband wireless network. It does this by encrypting connections between the SS and BS. In addition, the security sublayer provides operators with strong protection against theft of service. The BS protects against unauthorized access to this data transport service by enforcing encryption of the associated service flows across the network. The privacy sublayer employs an authenticated client/server key management protocol in which the BS, the server, controls distribution of keying material to its client SSs. Additionally, adding digital certificate-based SS authentication to its key management protocol strengthens the basic privacy mechanisms. Figure 7-1 illustrates the relationship of the MAC privacy layer with the MAC and physical layers.

Security Architecture in WiMAX   Privacy in the WiMAX specification has two component protocols:

- An encapsulation protocol for encrypting packet data across the fixed broadband wireless access (BWA) network. This protocol defines first a set of supported cryptographic suites (pairings of data encryption and authentication of algorithms and rules for applying those algorithms to a MAC PDU payload).

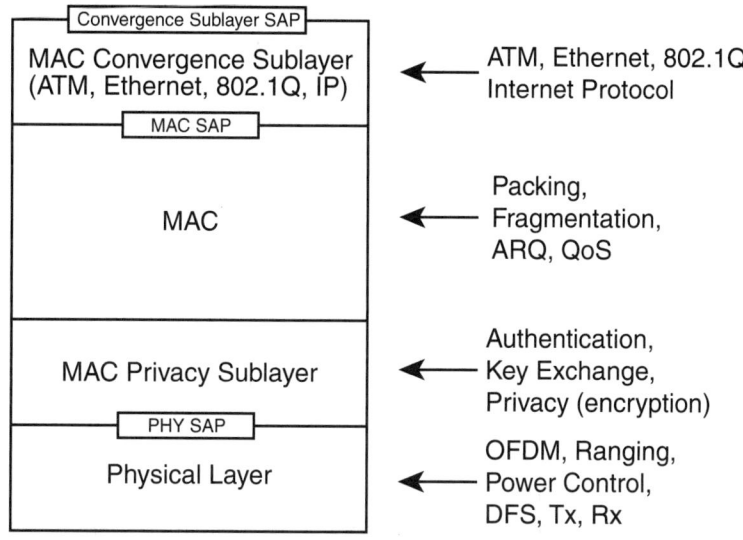

Figure 7-1
PHY and MAC
layers of WiMAX
specification
showing MAC
privacy sublayer
(Source: Intel)

■ A privacy key management (PKM) protocol providing the secure distribution of keying data from BS to SS. Through this key management protocol, the SS and the BS synchronize keying data; in addition, the BS uses the protocol to enforce conditional access to network services.

Packet Data Encryption   Encryption services are defined as a set of capabilities within the MAC security sublayer. MAC header information specific to encryption is allocated in the generic MAC header format. Encryption is always applied to the MAC PDU payload; the generic MAC header is not encrypted. All MAC management messages shall be sent in the clear to facilitate registration, ranging, and normal operation of the MAC.

Key Management Protocol   An SS uses the PKM protocol to obtain authorization and traffic keying material from the BS and to support periodic reauthorization and key refresh. The key management protocol uses X.509 digital certificates, the RSA public key encryption algorithm, and strong encryption algorithms to perform key exchanges between SS and BS.

The PKM protocol adheres to a client/server model where the SS (a PKM client) requests keying material, and the BS (a PKM server) responds to those requests. This protocol ensures that individual SS clients receive only keying material for which they are authorized. The PKM protocol uses MAC management messaging: PKM-REQ and PKM-RSP.

The PKM protocol uses public key cryptography to establish a shared, secret AK between the SS and the BS. The shared, secret key is then used to secure subsequent PKM exchanges of traffic encryption keys (TEKs). This two-tiered mechanism for key distribution permits refreshing of TEKs without incurring the overhead of computation-intensive public-key operations.

A BS authenticates a client SS during the initial authorization exchange. Each SS carries a unique X.509 digital certificate issued by the manufacturer. The digital certificate contains the SS's public key and MAC address. When requesting an AK, an SS presents its digital certificate to the BS. The BS verifies the digital certificate and then uses the verified public key to encrypt an AK that the BS then sends back to the requesting SS. Figure 7-2 details the relationship between X.509 and 56-bit DES.

The BS associates an SS's authentication identity with a paying subscriber and hence with the data service (voice, video, data) that subscriber is authorized to access. Thus, with the AK exchange, the BS establishes an authenticated identity of a client SS and the services the SS is authorized to access.

Figure 7-2
WiMAX security:
X.509
encryption and
56-bit DES

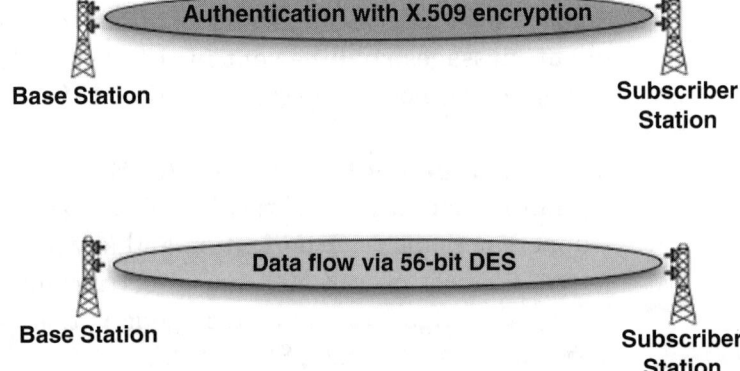

Because the BS authenticates the SS, it can protect against an attacker employing a cloned SS that is masquerading as a legitimate SS. The use of the X.509 certificates prevents cloned SSs from assigning fake credentials to a BS.

All SSs have factory-installed RSA private/public key pairs or provide an internal algorithm to generate such key pairs dynamically. If an SS relies on an internal algorithm to generate its RSA key pair, the SS shall generate the key pair prior to its first AK exchange. All SSs that rely on internal algorithms to generate an RSA key pair shall support a mechanism for installing a manufacturer-issued X.509 certificate following key generation.

The use of a factory-installed RSA private/public key pair limits the odds of success for any would-be hackers. The first hurdle for a would-be hacker is to have an SS from the same vendor as the targeted BS, and the second is to crack the X.509 encryption.[1]

In RSA, a message is encrypted with a public key and can only be decrypted with the corresponding private key. Any station can encrypt a message with the public key, but only one station can decrypt one using the secret private key. The flow of the encryption is *many to one*.

The reverse is also true: A message can be encrypted with a private key and can only be decrypted with the corresponding public key. This sort of *inside-out encryption*, to give it a name, might seem silly because anybody can use the public key to read the message. The flow of the encryption is *one to many*. Inside-out encryption provides no security, but the symmetry also holds, as the public key can *only* decrypt messages encrypted with the secret private key.[2]

Security Associations (SAs) An SA is the set of security information a BS and one or more of its client SSs share in order to support secure communications across the WiMAX standard. Three

---

[1]"802.16-2004 IEEE Standard for Local and Metropolitan Area Networks, Part 16, Air Interface for Fixed Broadband Wireless Access Systems," June 24, 2004, 271.

[2]Greg Goebel, 11.1 "Message Authentication & Digital Signatures" in *Codes, Ciphers & Codebreaking*, v.2.2.0, June 1, 2004, www.vectorsite.net/ttcodeb.html. p.1 section 11.1.

types of SAs are defined: Primary, Static, and Dynamic. Each manageable SS establishes a Primary Security association during the SS initialization process. Static SAs are provisioned within the BS. Dynamic SAs are established and eliminated on the fly in response to the initiation and termination of specific service flows. Both Static and Dynamic SAs can be shared by multiple SSs.

An SA's keying material has a limited lifetime. When the BS delivers SA keying material to an SS, it also provides the SS with that material's remaining lifetime. The SS is responsible for requesting new keying material from the BS before the set of keying material that the SS currently holds expires at the BS. Should the current keying material expire before a new set of keying material is received, the SS will perform network entry. The PKM protocol specifies how the SS and BS maintain key synchronizations.

The PKM Protocol

WiMAX utilizes PKM to establish a secure link between the base station and the subscriber station. The following paragraphs will describe this in greater detail.

SS Authorization and AK Exchange Overview   The SS authorization process includes the following steps:

- The BS authenticates a client SS's identity.
- The BS provides the authenticated SS with an AK from which a key encryption key (KEK) and message authentication keys are derived.
- The BS provides the authenticated SS with the identities and properties of primary and static SAs from which the SS is authorized to obtain keying information.

After achieving initial authorization, an SS periodically seeks reauthorization with the BS; reauthorization is also managed by the SS's authorization state machine. An SS must maintain its authorization status with the BS in order to be able to refresh aging TEKs.

An SS begins authorization by sending an Authentication Information message to its BS. The Authentication Information message contains the SS manufacturer's X.509 certificate, issued by the manufacturer itself or by an external authority.

The SS sends an Authorization Request message to its BS immediately after sending the Authentication Information message. This is a request for an AK, as well as for the Security Association Identifications (SAIDs) identifying any Static Security SAs the SS is authorized to participate in. The Authorization Request message includes the following:

- A manufacturer-issued X.509 certificate

- A description of the cryptographic algorithms the requesting SS supports; an SS's cryptographic capabilities are presented to the BS as a list of cryptographic suite identifiers, each indicating a particular pairing of packet data encryption and packet data authentication algorithms the SS supports

- The SS's Basic CID

In response to an Authorization Request message, a BS validates the requesting SS's identity, determines the encryption algorithm and protocol support it shares with the SS, activates an AK for the SS, encrypts it with the SS's public key, and sends it back to the SS in an Authorization Reply message. The authorization reply includes the following:

- An AK encrypted with the SS's public key

- A four-bit key sequence number used to distinguish between successive generations of AKs

- A key lifetime

- The identities and properties of the single primary and zero or more static SAs for which the SS is authorized to obtain keying information

In responding to an SS's Authorization Request, the BS shall determine whether the requesting SS, whose identity can be verified via the X.509 digital certificate, is authorized for basic unicast services and what additional statically provisioned services the SS's user has subscribed for.

An SS periodically refreshes its AK by reissuing an Authorization Request to the BS. Reauthorization is identical to authorization with the exception that the SS does not send Authentication Information messages during reauthorization cycles.

TEK Exchange Overview

Upon receiving authorization, an SS starts a separate TEK state machine for each of the SAIDs identified in the Authorization Reply message. Each TEK state machine operating within the SS is responsible for managing the keying material associated with its respective SAID. TEK state machines periodically send Key Request messages to the BS, requesting a refresh of keying material for their respective SAIDs.

The BS responds to a Key Request with a Key Reply message containing the BS's active keying material for a specific SAID. The TEK is encrypted using KEK derived from the AK.

The Key Reply provides the requesting SS the remaining lifetime of each of the two sets of keying material. The receiving SS uses these remaining lifetimes to estimate when the BS will invalidate a particular TEK and, therefore, when to schedule future Key Requests so that the SS requests and receives new keying material before the BS expires the keying material the SS currently holds.[3] Table 7-1 details this process.

Cryptographic Methods

Once the authentication process is complete, the next step is for the data flow to be encrypted. The following sections will describe this process.

---

[3]"Air Interface for Fixed Broadband Wireless Access Systems," 272–275.

Table 7-1

PKM Exchange
Messages

PKM Message	Description
Authentication Information	Contains the manufacturer's X.509 certificate (issued by an external authority)
Authorization Request	Sent from an SS to its BS to request an AK and list of authorized SAIDs
Authorization Reply	Sent from a BS to an SS to reply to an AK and a list of authorized SAIDs
Authorization Invalid	Sent from a BS to an SS to reject an Authorization Request message received from that SS
Key Request	Sent from an SS to its BS to request a TEK for the privacy of one of its authorized SAIDs
Key Reply	Sent from a BS to an SS to carry the two active sets of traffic keying material for the SAID
Key Reject	Sent from a BS to an SS to indicate that the SAID is no longer valid and no key will be sent
TEK Invalid	Sent from a BS to an SS if it determines that the SS encrypted the UL with an invalid TEK
SA Add	Sent from a BS to an SS to establish one or more SAs

Source: IEEE

Data Encryption with DES in CBC Mode If the data encryption algorithm identifier in the cryptographic suite of an SA equals 0x01, data on connections associated with the SA shall use the CBC mode of the United States Data Encryption Standard (DES) algorithm to encrypt the MAC PDU payloads.

The CBC IV shall be calculated as follows: in the DL, the CBC shall be initialized with the exclusive-or (XOR) of (a) the IV parameter included in the TEK keying information and (b) the content of the

PHY Synchronization field of the latest DL-MAP. In the UL, the CBC shall be initialized with the XOR of (a) the IV parameter included in the TEK keying information and (b) the content of the PHY Synchronization field of the DL-MAP that is in effect when the UL-MAP for the UL transmission is created/received.

Residual termination block processing shall be used to encrypt the final block of plaintext when the final block is less than 64 bits. Given a final block having n bits, where n is less than 64, the next-to-last ciphertext block shall be DES encrypted a second time, using the electronic code book (ECB) mode, and the most significant n bits of the result are XORed with the final n bits of the payload to generate the short final cipherblock. In order for the receiver to decrypt the short final cipherblock, the receiver DES encrypts the next-to-last ciphertext block, using the ECB mode, and XORs the most significant n bits with the short final cipher block in order to recover the short final clear text block.[4]

Conclusion

This chapter covered the security mechanisms built into the IEEE 802.16 WiMAX specification. It is encouraging to note that, unlike its 802.11 predecessors, WiMAX has powerful security measures at its launch. Although there is no such thing as an unhackable network, the incorporation of the two-stage security process (X.509 in the authentication process and 56-bit DES for the service flow) will deter all but the most dedicated and knowledgeable hackers.

---

[4]"Air Interface for Fixed Broadband Wireless Access Systems," 295.

WiMAX VoIP

Telephone companies are threatened because it is infinitely cheaper to beam data (and voice) to a customer than it is to run a copper wire or coax cable to them. In addition, the potential data flow to a subscriber over a WiMAX network is exponentially greater than the 56 Kbps delivered via a telco's copper wire dial-up connection. The emergence of softswitch as a switching alternative to Class 4 and Class 5 switches makes it all the more feasible for WiMAX service providers to offer voice services independent of the telephone company or for subscribers (especially enterprises) to be their own telephone company, effectively bypassing the PSTN entirely.

PSTN Architecture

The PSTN, over which the vast majority of the voice traffic in North America travels, is comprised of three elements: transport, the transportation of conversation from one CO to another; switching, the switching or routing of calls in the PSTN via a telephone switch contained in the CO; and access, the connection between the switch in the CO and the subscriber's telephone or other telecommunications device. Figure 8-1 provides an overview of this architecture.

Figure 8-1
The three components of the PSTN— access, switching, and transport—and their WiMAX counterparts

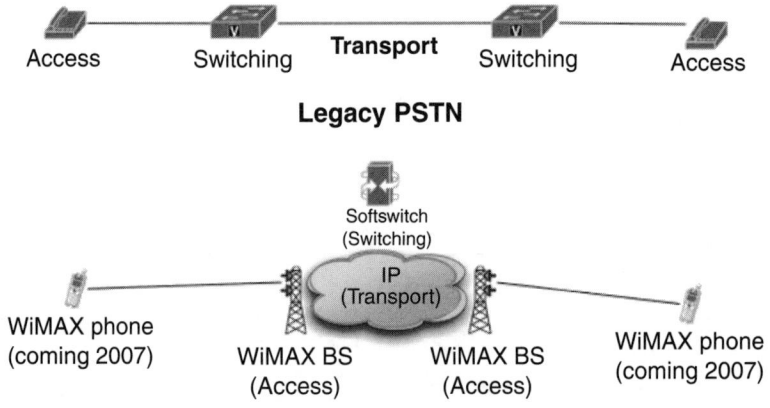

Legacy PSTN

Access   Switching   **Transport**   Switching   Access

Softswitch (Switching)

IP (Transport)

WiMAX phone (coming 2007)   WiMAX BS (Access)   WiMAX BS (Access)   WiMAX phone (coming 2007)

PSTN Bypass with WiMAX and VoIP

As illustrated in Figure 8-1, WiMAX is a form of access to a wider network (PSTN, corporate LAN or WAN, or Internet). The MFJ of 1984 opened transport to competition. The *bandwidth glut* currently has made transport via IP backbone relatively inexpensive. The use of WiMAX as a backhaul mechanism will only accelerate that trend. Softswitch technologies (IP PBX, Class 4 and 5 replacements) offer a viable alternative to the switching facilities of the PSTN. The Telecommunications Act of 1996 was intended to open the switching and access facilities of the PSTN to competition. For a number of reasons, this has not happened. WiMAX presents a bypass technology of the telco's copper wire access.

Voice Over WiMAX—The Challenge

The emerging popularity of VoIP in the enterprise market coupled with WiMAX raises the question: Can voice be transported over a WiMAX network? This chapter will discuss the objections to transmitting voice over WiMAX networks and will offer solutions to those objections.

VoIP

The emergence of VoIP raises a wide range of possibilities. By virtue of transporting voice over a data stream, VoIP frees the voice stream from the confines of a voice-specific network and its associated platforms. VoIP can be received and transmitted via PCs, laptops, IP, and Wi-Fi handsets. Where there is IP, there can be VoIP.

Origins of VoIP

In November 1988, Republic Telcom (yes, one "e") of Boulder, Colorado, received patent number 4,782,485 for a Multiplexed Digital Packet Telephone System. The plaque from the Patent and Trademark Office describes it as follows:

A method for communicating speech signals from a first location to a second location over a digital communication medium comprising the steps of: providing a speech signal of predetermined bandwidth in analog signal format at said first location; periodically sampling said speech signal at a predetermined sampling rate to provide a succession of analog signal samples; representing said analog signal samples in a digital format thereby providing a succession of binary digital samples; dividing said succession of binary digital samples into groups of binary digital samples arranged in a temporal sequence; transforming at least two of said groups of binary digital samples into corresponding frames of digital compression.

Republic and its acquiring company, Netrix Corporation, applied this voice over data technology to the data technologies of the times (X.25 and frame relay) until 1998 when Netrix and other competitors introduced VoIP onto their existing voice over data gateways. While attempts had been made at Internet telephony from a software-only perspective, commercial applications were limited to using voice over data gateways that could interface the PSTN to data networks. Voice over data applications were popular in enterprise networks with offices spread across the globe (eliminating international interoffice long-distance bills), in offices where no PSTN existed (installations for mining and oil companies), and for long-distance bypass (legitimate and illegitimate).

How Does VoIP Work?

The first process in an IP voice system is the digitization of the speaker's voice. The next step (and the first step when the user is on a handset connected to a gateway using a digital PSTN connection) is typically the suppression of unwanted signals and compression of the voice signal. This step has two stages. First, the system examines the recently digitized information to determine if it contains voice signal or only ambient noise and discards any packets that do not contain speech. Second, complex algorithms are employed to reduce the amount of information that must be sent to the other party.

Sophisticated codecs enable noise suppression and compression of voice streams. Compression algorithms (also known as codecs or coders/decoders) include G.723, G.728, and G.729. G.711 is the codec for uncompressed voice at 64 Kbps.

Following compression, voice must be packetized and VoIP signaling protocols added. Some storage of data occurs during the process of collecting voice data because the transmitter must wait for a certain amount of voice data to be collected before it is combined to form a packet and transmitted via the network. Protocols are added to the packet to facilitate its transmission across the network. For example, each packet will need to contain the address of its destination, a sequencing number in case the packets do not arrive in the proper order, and additional data for error checking.

Because IP is a protocol designed to interconnect networks of varying kinds, substantially more processing is required than in smaller networks. The network-addressing system can often be very complex, requiring a process of encapsulating one packet inside another and, as data moves along, repackaging, readdressing, and reassembling the data.

When each packet arrives at the destination computer, its sequencing is checked to place the packets in the proper order. A decompression algorithm is used to restore the data to its original form, and clock-synchronization and delay-handling techniques are used to ensure proper spacing. Because data packets are transported via the network by a variety of routes, they do not arrive at their destination in order. To correct this situation, incoming packets are stored for a time in a jitter buffer to wait for late-arriving packets. The length of time in which data are held in the jitter buffer varies, depending on the characteristics of the network.

VoIP Signaling Protocols

VoIP signaling protocols, H.323 and SIP, set up the route for the media stream or conversation over an IP network. Gateway control protocols, such as Media Gateway Control Protocol (MGCP), and signaling protocols establish control and status in media and signaling gateways.

Once the route of the media stream has been established, routing (User Diagram Protocol [UDP] and Transmission Control Protocol [TCP]) and transporting (Real-Time Transport Protocol [RTP]) the media stream (conversation) are the function of routing and transport protocols. Routing protocols, such as UDP and TCP, could be compared to the *switching* function described in Chapters 2 and 3.

RTP would be analogous to the *transport* function in the PSTN. The signaling and routing functions establish what route the media stream will take when the routing protocols delivers the bits, that is, the conversation.

Setting up a VoIP call is roughly similar to setting up a circuit-switched call on the PSTN. A media gateway or IP phone must be loaded with the parameters to allow proper media encoding and the use of telephony features. Inside the media gateway is an intelligent entity known as an endpoint. When the calling and called parties agree on how to communicate and the signaling criteria is established, the media stream over which the packetized voice conversation will flow is established. Signaling establishes the virtual circuit over the network for that media stream. Signaling is independent of the media flow. It determines the type of media to be used in a call and is concurrent throughout the call. Two types of signaling are currently popular in VoIP: H.323 and Session Initiation Protocol (SIP).[1]

Figure 8-2 details the relationship between signaling and media flow. VoIP's relationship between transport and signaling resembles PSTN's, in that SS7 is out-of-channel signaling, as is used in VoIP.

H.323 H.323 is the International Telecommunications Union (ITU-T) recommendation for packet-based multimedia communication. H.323 was developed before the emergence of VoIP for video over a local area network (LAN). As it was not specifically designed for VoIP, H.323 has faced a good deal of competition from a competing protocol, SIP, which was designed specifically for VoIP over any size of network. H.323 has enjoyed a first mover advantage, and there now exists a considerably installed base of H.323 VoIP networks.

---

[1]Bill Douskalis, *IP Telephony: The Integration of Robust VoIP Services* (Upper Saddle River, NJ: Prentice Hall, 2000).

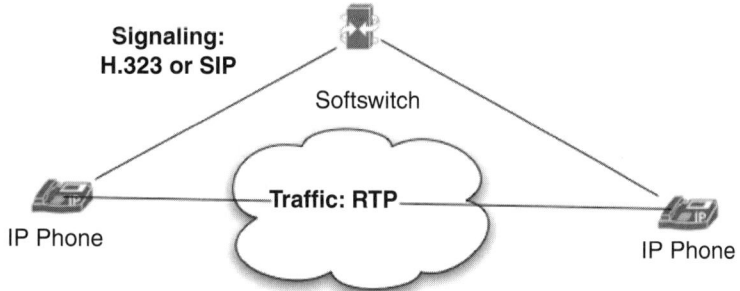

Figure 8-2
Signaling and transport protocols used in VoIP

H.323 is made up of a number of subprotocols. It uses protocol H.225.0 for registration, admission, status, call signaling, and control. It also uses protocol H.245 for media description and control, terminal capability exchange, and general control of the logical channel carrying the media stream(s). Other protocols make up the complete H.323 specification, which presents a protocol stack for H.323 signaling and media transport. H.323 also defines a set of call control, channel setup, and codec specifications for transmitting real-time video and voice over networks that don't offer guaranteed service or quality of service. As a transport, H.323 uses RTP, an Internet Engineering Task Force (IETF) standard designed to handle the requirements of streaming real-time audio and video via the Internet.[2]

SIP: Alternative Softswitch Architecture? If the worldwide PSTN could be replaced overnight, the best candidate architecture at this time would be based on VoIP and SIP. Much of the VoIP industry has been based on offering solutions that leverage existing circuit-switched infrastructure (for example, VoIP gateways that interface a private branch exchange [PBX] and an IP network). At best, these solutions offer a compromise between circuit- and packet-switching architectures with resulting liabilities of limited features, expensive-to-maintain circuit-switched gear, and questionable QoS and reliability, as a call is routed between networks based on those technologies. SIP is an architecture that potentially offers more features than a circuit-switched network.

[2]Ibid., 9.

SIP is a signaling protocol. It uses a text-based syntax similar to Hypertext Transfer Protocol (HTTP), as used in web addresses. Programs that are designed for parsing of HTTP can be adapted easily for use with SIP. SIP addresses, known as SIP URLs (or uniform resource locators), take the form of web addresses. A web address can be the equivalent of a telephone number in an SIP network. In addition, PSTN phone numbers can be incorporated into an SIP address for interfacing with the PSTN. An e-mail address is portable. Using the proxy concept, one can check his or her e-mail from any Internet-connected terminal in the world. Telephone numbers, simply put, are not portable; they ring at only one physical location. SIP offers a mobility function that can follow subscribers to the nearest phone at a given time.

Like H.323, SIP handles the setup, modification, and teardown of multimedia sessions, including voice. While it works with most transport protocols, its optimal transport protocol is RTP. Figure 8-2 shows how SIP functions as a signaling protocol, while RTP is the transport protocol for a voice conversation. SIP was designed as a part of the IETF multimedia data and control architecture. It is designed to interwork with other IETF protocols such as Session Description Protocol (SDP), RTP, and Session Announcement Protocol (SAP). It is described in the IETF's RFC 2543. Many in the VoIP and softswitch industry believe that SIP will replace H.323 as the standard signaling protocol for VoIP.

SIP is part of the IETF standards process and is modeled on other Internet protocols such as Simple Mail Transfer Protocol (SMTP) and HTTP. It is used to establish, change, and tear down (end) calls between one or more users in an IP-based network. In order to provide telephony services, a number of different standards and protocols need to come together—specifically to ensure transport (RTP) signaling with the PSTN, guarantee voice quality (Resource Reservation Protocol [RSVP]), provide directories (Lightweight Directory Access Protocol [LDAP]), authenticate users (remote authentication dial-in user service [RADIUS]), and scale to meet anticipated growth curves.

How Does SIP Work?   SIP is focused on two classes of network entities: clients (also called user agents [UAs]) and servers. VoIP calls on SIP to originate at a client and terminate at a server. Types of clients in the technology currently available for SIP telephony include a personal computer (PC) loaded with a telephony agent or a SIP telephone. Clients can also reside on the same platform as a server. For example, a PC on a corporate WAN might be the server for the SIP telephony application, but it might also function as a user's telephone (client).

SIP Architecture   SIP is a client-server architecture. The client in this architecture is the UA. The UA interacts with the user. It usually has an interface toward the user in the form of a PC or an IP phone (SIP phone in this case). There are four types of SIP servers: UA server, redirect server, proxy server, and a registrar. The type of SIP server used determines the architecture of the network.

Switching

In the PSTN, the switching function is performed in the CO, which contains a Class 5 switch for local calls and a Class 4 switch for long-distance calls. A Class 5 switch can cost upwards of tens of millions of dollars and is very expensive to maintain. This expense has kept competitors out of the local calling market. A new technology known as softswitch is far less expensive in terms of purchase and maintenance. Potentially, softswitch allows a competitive service provider to offer their own service without having to route calls through the incumbent service provider's CO. The following pages describe softswitch.

Softswitch (aka Gatekeeper, Media Gateway Controller)

In a VoIP network, a softswitch is the intelligence that coordinates call control, signaling, and features that make a call across a network or multiple networks possible. A softswitch primarily performs call control (call set-ups and teardowns). Once a call is set up, connection control ensures that the call stays up until the originating or terminating user releases it. Call control and service logic refer to the functions that process a call and offer telephone features. Examples of call control and service logic functions include recognizing that a party has gone off hook and that a dial tone should be provided, interpreting the dialed digits to determine where the call is to be terminated, determining if the called party is available or busy, and finally, recognizing when the called party answers the phone and when either party subsequently hangs up and then recording these actions for billing.

A softswitch coordinates the routing of signaling messages between networks. Signaling coordinates actions associated with a connection to the entity at the other end of the connection. To set up a call, a common protocol must be used that defines the information in the messages and that is intelligible at each end of the network and across dissimilar networks. The main types of signaling a softswitch performs are peer-to-peer for call control and softswitch-to-gateway for media control. For signaling, the predominant protocols are SIP, Signaling System 7 (SS7), and H.323. For media control, the predominant signaling protocol is MGCP.

As a point of introduction to softswitch, it is necessary to clarify the evolution to softswitch and define media gateway controller and gatekeeper, the precursors to softswitch. Media Gateway Controllers (MGC) and gatekeepers (essentially synonymous terms for the earliest forms of softswitch) were designed to manage low-density (relative to a carrier grade solution) voice networks. MGC communicates with both the signaling gateway and the media gateway to provide the necessary call processing functions. The MGC uses either MGCP or MEGACO/H.248 (described in a later chapter) for intergateway communications.

Gatekeeper technology evolved out of H.323 technology (a VoIP signaling protocol described in the next chapter). As H.323 was designed for LANs, an H.323 gatekeeper can only manage activities in a zone (read LAN but not specifically an LAN). A zone is a collection of one or more gateways managed by a single gatekeeper. A gatekeeper should be thought of as a logical function, not a physical entity. The functions of a gatekeeper are address translation (that is, a name or e-mail address for a terminal or gateway and a transport address) and admissions control (authorizes access to the network).

As VoIP networks got larger and more complex, management solutions with far greater intelligence became necessary. Greater call processing power was needed, as was the ability to interface signaling between IP networks with the PSTN (VoIP signaling protocols to SS7). Other drivers included the need to integrate features on the network and interface disparate VoIP protocols. Thus was born the softswitch.

The softswitch provides usage statistics to coordinate billing and to track operations and administrative functions of the platform while interfacing with an application server to deliver value-added subscriber services. The softswitch controls the number and type of features provided. It interfaces with the feature/application server to coordinate features (conferencing, call forwarding, and so on) for a call.

Physically, a softswitch is software hosted on a server chassis filled with IP boards and includes the call control applications and drivers.[3] Very simply, the more powerful the server, the more capable the softswitch. That server need not be colocated with other components of the softswitch architecture.

Other Softswitch Components

Softswitch's key advantage over its circuit-switched predecessor is utilizing distributed architecture. That is, its components need not be colocated. Those components include signaling gateway, media gateway, and application server.

Signaling Gateway Signaling gateways are used to terminate signaling links from PSTN networks or other signaling points. The SS7 signaling gateway serves as a protocol mediator (translator) between the PSTN and IP networks. That is, when a call originates in an IP network using H.323 as a VoIP protocol and must terminate in the PSTN, a translation from the H.323 signaling protocol to SS7 is necessary in order to complete the call. Physically, the signaling

---

[3]Ibid.

function can be embedded directly into the media gateway controller or housed within a stand-alone gateway.

Media Gateway The media gateway converts an analog or circuit-switched voice stream to a packetized voice stream. Media gateways rank from one or two port residential gateways to carrier-grade platforms with 100,000 ports. The media gateway can be located at the customer's premises or colocated at the CO.

Application Server The application server accommodates the service and feature applications made available to the service provider's customers. These applications include call forwarding, conferencing, voice mail, forward on busy, and so on. Physically, an application server is a server loaded with a software suite that offers the application programs. The softswitch accesses these, then enables and applies them to the appropriate subscribers as needed. Figure 8-3 illustrates the relationship of softswitch components.

A softswitch solution emphasizes open standards, as opposed to the Class 4 or 5 switch that historically offered a proprietary and closed environment. A carrier was a "Nortel shop" or a "Lucent shop." No components (hardware or software) from one vendor were compatible with products from another vendor. Any application or feature on a DMS-250, for example, had to be a Nortel product or specifically approved by Nortel. This usually translated to less than

Figure 8-3
Relationship of softswitch components

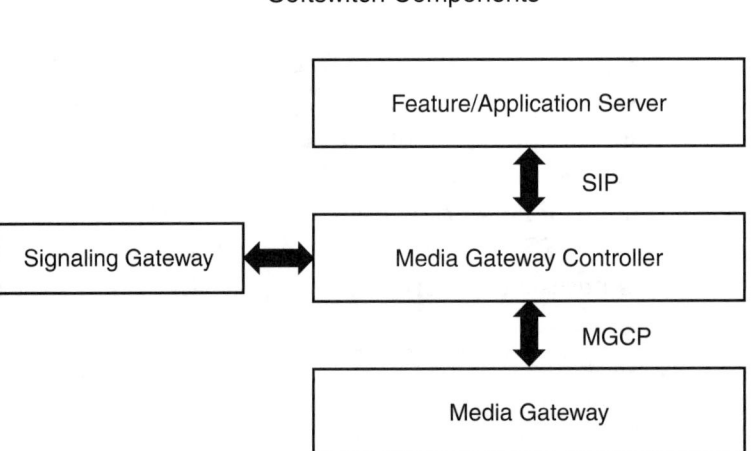

Softswitch Components

Feature/Application Server

SIP

Signaling Gateway ⟷ Media Gateway Controller

MGCP

Media Gateway

competitive pricing for those components. Softswitch open standards are aimed at freeing service providers from vendor dependence and the long and expensive service development cycles of legacy switch manufacturers.

Service providers are concerned with whether a softswitch solution can transmit a robust feature list identical to that found on a 5ESS Class 5 switch, for example. Softswitch offers the advantage of allowing a service provider to integrate third-party applications or even write its own while interoperating with the features of the PSTN via SS7. This is potentially the greatest advantage to a service provider presented by softswitch technology.

Features reside at the application layer in softswitch architecture. The interface between the Call Control Layer and specific applications is Application Program Interface (API). Writing and interfacing an application with the rest of the softswitch architecture occurs in the Service Creation Environment.

VoIP and Softswitch Pave the Way for Voice Over WiMAX

A number of attempts have been made to deploy voice services via wireless local loop (WLL): for instance, using wireless technologies (not WiMAX) to offer telephone service in underserved or third-world markets. As these services have been limited to voice services, they have not caught on with a mass market. WiMAX is different, as it is a protocol for Ethernet over a wireless medium. The building blocks for a potential alternative to the PSTN now fall into place. Not only does a WiMAX network offer a potential bypass of the PSTN for voice services, it also offers broadband Internet and its incumbent suite of services.

Objections to VoIP Over WiMAX

Like concerns over WiMAX as a whole, five major objections arise to adopting VoIP over WiMAX: voice quality as it relates to QoS, security, E911, Communications Assistance to Law Enforcement Agencies

(CALEA), and range. Many suspect that there is an adverse trade-off between the predictability (QoS) of copper wires that deliver voice to residences and the unpredictability of the airwaves as utilized by WiMAX.

Objection One: Voice Quality of WiMAX VoIP

Despite the fact that telephone companies in the United States are losing thousands of lines per month to cell phone service providers, many believe that voice over a cell phone connection would deliver inferior voice quality and, as a result, is not a viable alternative to the copper wires of the PSTN.

Measuring Voice Quality in WiMAX VoIP   How does one measure the difference in voice quality between a WiMAX VoIP and the PSTN? As the VoIP industry matures, new means of measuring voice quality are coming on the market. Currently, two tests award some semblance of a score for voice quality. The first, Mean Opinion Score (MOS), is a holdover from the circuit-switched voice industry, and the other, Perceptual Speech Quality Measurement (PSQM), has emerged with VoIP's increasing popularity.

Mean Opinion Score (MOS)   Can voice quality as a function of QoS be measured scientifically? The telephone industry employs a subjective rating system known as the MOS to measure the quality of the telephone connections. The measurement techniques are defined in ITU-T P.800 and are based on the opinions of many testing volunteers who listen to a sample of voice traffic and rate the quality of that transmission. The volunteers listen to a variety of voice samples and are asked to consider factors such as loss, circuit noise, side tone, talker echo, distortion, delay, and other transmission problems. The volunteers then rate the voice samples from 1 to 5, with 5 being "Excellent" and 1 being "Bad." The voice samples are then awarded a Mean Opinion Score, or MOS. A MOS of 4 is considered "toll quality."

It should be stated here that the voice quality of VoIP applications can be engineered to be as good or better than the PSTN. Recent research performed by the Institute for Telecommunications Sciences in Boulder, Colorado, compared the voice quality of traffic routed through VoIP gateways with the PSTN. Researchers were fed a variety of voice samples and were asked to determine if the sample originated with the PSTN or from the VoIP gateway traffic. The test indicated that the voice quality of the VoIP gateway routed traffic was "indistinguishable from the PSTN."[4] It should be noted that the IP network used in this test was a closed network, not the public Internet or other long-distance IP network. This report indicates that quality media gateways can deliver voice quality on the same level as the PSTN. The challenge then shifts to ensuring the IP network can deliver similar QoS to ensure good voice quality. This chapter will explain how measures can be taken to engineer voice-specific solutions into a wireless network to ensure voice quality equals the PSTN.

Perceptual Speech Quality Measurement (PSQM) Another means of testing voice quality is known as PSQM. This method, based on ITU-T Recommendation P.861, specifies a model to map actual audio signals to their representations inside the head of a human. Voice quality consists of a mix of objective and subjective parts; it varies widely among the different coding schemes and the types of network topologies used for transport. In PSQM, measurements of processed (compressed, encoded, and so on) signals derived from a speech sample are collected, an objective analysis is performed comparing the original and the processed version of the speech sample, and an opinion as to the quality of the signal processing functions that processed the original signal is rendered. Unlike MOS scores, PSQM scores result in an absolute number, not a relative comparison between the two signals.[5] This is valuable because vendors can state the PSQM score for a given platform (as assigned by an impartial testing agency). Service providers can then make at least part of their buying decision based on the PSQM score of the platform.

---

[4]Andrew Craig, "Qualms of Quality Dog Growth of IP Telephony," *Network News* (November 11, 1999): 3.

[5]Bill Douskalis, *IP Telephony*, 242–243.

Detractors to Voice Quality in WiMAX What specifically detracts from good voice quality in a WiMAX environment? Latency, jitter, packet loss, and echo are the problems. With proper engineering, the impact of these factors on voice quality can be minimized, and voice quality equal to or better than the PSTN can be achieved on WiMAX networks.

Latency (aka Delay) Voice as a wireless IP application presents unique challenges for WiMAX networks. Primary among these is acceptable audio quality resulting from minimized network delay in a mixed voice and data environment. Ethernet, wired or wireless, was not designed for real-time streaming media or guaranteed packet delivery. Congestion without traffic differentiation on the wireless network can quickly render voice unusable. QoS measures must be taken to ensure voice packet delays remain under 100 ms.

Voice signal processing at the sending and receiving ends, which includes the time required to encode or decode the voice signal from analog or digital form into the voice-coding scheme selected for the call and vice versa, adds to the delay. Compressing the voice signal will also increase the delay: the greater the compression, the greater the delay. Where bandwidth costs are not a concern, a service provider can utilize G.711, uncompressed voice (64 Kbps), which imposes a minimum of delay due to the lack of compression.

On the transmit side, packetization delay is another factor that must be entered into the calculations. The packetization delay is the time it takes to fill a packet with data: the larger the packet size, the more time required. Using smaller packet sizes can shorten this delay but will increase the overhead because more packets containing similar information in the header have to be sent. Balancing voice quality, packetization delay, and bandwidth utilization efficiency is very important to the service provider.[6]

How much delay is too much? Of all the factors that degrade VoIP, latency (or delay) is the greatest problem. Recent testing by Mier Labs offers a metric as to how much latency is acceptable or comparable to toll quality, the voice quality offered by the PSTN. Latency

---

[6]Ibid., 230–231.

less than 100 ms does not affect toll-quality voice. However, latency over 120 ms is discernable to most callers, and at 150 ms the voice quality is noticeably impaired, resulting in less than toll-quality communication. The challenge for VoIP service providers and their vendors is to keep the latency of any conversation on their network from exceeding 100 ms.[7] Humans are intolerant of speech delays of more than about 200 ms. As mentioned earlier, ITU-T G.114 specifies that delay is not to exceed 150 ms one-way or 300 ms round trip. The dilemma is that while elastic applications (e-mail, for example) can tolerate a fair amount of delay, they usually try to consume every possible bit of network capacity. In contrast, voice applications need only a small amount of the network, but that amount has to be available immediately.[8]

Dropped Packets   In IP networks, a percentage of the packets can be lost or delayed, especially in periods of congestion. Also, some packets are discarded due to errors that occurred during transmission. Lost, delayed, and damaged packets result in substantial deterioration of voice quality. In conventional error correction techniques used in other protocols, incoming blocks of data containing errors are discarded, and the receiving computer requests the retransmission of the packet; thus, the message that is finally delivered to the user is exactly the same as the message that originated.

As VoIP (and tangentially WiMAX VoIP) systems are time sensitive and cannot wait for retransmission, more sophisticated error detection and correction systems are used to create sound to fill in the gaps. This process stores a portion of the incoming speaker's voice; then, using a complex algorithm to approximate the contents of the missing packets, new sound information is created to enhance the communication. Thus, the sound the receiver hears is not exactly the sound transmitted; rather, portions have been created by the system to enhance the delivered sound.[9]

---

[7]Mier Communications, "Lab Report—QoS Solutions," February 2001, p. 2, www.sitaranetworks.com/solutions/pdfs/mier_report.pdf.

[8]John McCullough and Daniel Walker, "Interested in VoIP? How to Proceed," *Business Communications Review* (April 1999): 16–22.

[9]Report to Congress on Universal Service, CC Docket No. 96–45, white paper on IP Voice Services, March 18, 1998, www.von.org/docs/whitepap.pdf.

Jitter Jitter occurs because packets have varying transmission times. It is caused by different queuing times in the routers and possible different routing paths. Jitter results in unequal time spacing between the arriving packets and requires a jitter buffer to ensure a smooth, continuous playback of the voice stream.

The chief correction for jitter is to include an adaptive jitter buffer. An adaptive jitter buffer can dynamically adjust to accommodate for high levels of delay that can be encountered in wireless networks.

A Word About Bit Rate (or Compression Rate) The bit rate, which is the number of bits per second delivered by the speech encoder, determines the bandwidth load on the network. It is important to note that the packet headers (IP, UDP, and RTP) also add to the bandwidth. Speech quality generally increases with the bit rate: Very simply put, the greater the bandwidth, the greater the speech quality.

Solution: Voice Codecs Designed for VoIP, Especially VoIP Over WiMAX

Many of the detractors to good speech quality in VoIP over WiMAX can be overcome by engineering a variety of fixes into the speech codecs used in both circuit- and packet-switched telephony. The following sections will describe speech coding and explain how it applies to speech quality.

Modifying Voice Codecs to Improve Voice Quality

One of the first processes in the transmission of a telephone call is the conversion of an analog signal (the wave of the voice entering the telephone) into a digital signal. This process is called pulse code modulation (PCM). PCM is a four-step process consisting of pulse amplitude modulation (PAM) sampling, companding, quantization, and encoding. Encoding is a critical process in VoIP and WiMAX VoIP. To date, voice codecs used in VoIP (packet switching) are taken directly from PSTN technologies (circuit switching). Cell phone technologies use PSTN voice codecs. New software in the WiMAX VoIP industry utilizes modified PSTN codecs to deliver voice quality comparable to the PSTN.

Popular Speech Codecs Speech codecs, based largely on compression algorithms, are a significant determinant in the quality of a telephone conversation.

The QoS Solution: Fix Circuit-Switched Voice Codecs in a Packet Switched, Wireless World with Enhanced Speech-Processing Software

If circuit-switching voice codecs are the challenge to good QoS in wireless, packet-switched networks, what, then, is the fix for outdated voice codecs? An emerging market of enhanced speech-processing software corrects for the shortcomings of traditional voice codecs, which were designed decades ago for a circuit-switched PSTN. These recent developments in VoIP software provide QoS enhancement solutions for IP telephony in the terminal with very high voice quality even with severe network degradations caused by jitter and packet loss. These VoIP QoS enhancements should provide WiMAX VoIP speech quality comparable to that of the PSTN. Also, speech quality should degrade gradually as packet loss increases. Moderate packet loss percentages should be inaudible. Table 8-1 shows the relationship between speech codecs and MOS scores.

Enhanced Speech-Processing Software New speech-processing algorithms provide for diversity; this means that an entire speech segment is not lost when a single packet is lost. Diversity is achieved by reorganizing the representation of the speech signal. Diversity

Table 8-1

MOS Scores of
Speech Codecs

Standard	Data Rate (Kbps)	Delay (ms)	MOS
G.711	64	0.125	4.8
G.721			
G.723			
G.726	16, 24, 32, 40	0.125	4.2
G.728	16	2.5	4.2
G.729	8	10	4.2
G.723.1	5.3, 6.3	30	3.5, 3.98

does not add redundancy or send the same information twice. There-fore, it is bandwidth efficient and ensures that packet losses lead to a gradual and imperceptible degradation of voice quality. The trade-off is that diversity leads to increased delays. Enhanced speech-processing software includes advanced signal processing to dynamically minimize delay. Therefore, the overall delay is main-tained at approximately the same level that it would be without diversity. Furthermore, the basic quality (no packet loss) is equiva-lent to or better than PSTN (using G.711).

Enhanced speech-processing software is built to enhance existing standards used in IP telephony. This software enables high speech quality on a loaded network with jitter, high packet losses, and delays. Cost savings are realized using enhanced speech-processing software, as there is no need to overprovision network infrastructure. The high packet loss tolerance also reduces the need for and subse-quent cost of network supervision, resulting in further cost savings.

Objection Two: Security for WiMAX VoIP

Although Chapter 7 was devoted to security of WiMAX networks, it is important to examine how security applies specifically to voice over WiMAX. A misperception is that the voice stream is susceptible to interception (eavesdropping) because the conversation is transmitted over the airwaves. Given the double encryption process of WiMAX (X.509 and 56-bit DES), it would be extremely difficult to tap into such a conversation.

Objection Three: CALEA and E911

The circuit-switched industry's common objection to VoIP concerns a telecommunications carrier's compliance with CALEA and E911, the legal requirements for primary line telephone service providers in the United States. The laws requiring telephone companies to provide these services were made before the Internet came to mainstream America. Although there are technological means of providing these services in a number of circumstances, the first question should concern the service provider's obligation in providing these services.

E911

A number of E911 solutions are coming on the market at the time of this writing. First, some solutions overflow from the cell phone industry where E911 will soon be a requirement. Another solution is to include global positioning satellite (GPS) technology in a WiMAX VoIP handset. That way, the exact location of the handset would be known at any time.

More concretely, E911 compliance is possible by registering the IP telephone with the PSTN's PSAP database, which maps a telephone number to a physical location. For those who use their IP telephone as a static home or office phone, this may present a suitable solution. For those who take their IP phone traveling, it becomes necessary to reregister that handset's location upon being installed at the new location.

Communications Assistance to Law Enforcement Agencies (CALEA) This requirement may be relaxed in a forthcoming regulatory regime outlined by former FCC Chairman Michael Powell in an October 30, 2002 address at the University of Colorado. In that speech, Chairman Powell conceded that the CALEA law was designed for the circuit-switched world and is (at the time of that speech) difficult to comply with in a WiMAX VoIP environment. As a result, and in the interest of promoting all that WiMAX and similar technologies have to offer, Chairman Powell hinted that such requirements would have to be relaxed.

Very simply put, CALEA calls for two capabilities. First, it calls for the collection of call details: who called whom, when, and for how long. This is not difficult in a VoIP environment, as most softswitch products can collect call detail records (CDRs). Second, and this is the hard part, CALEA calls for the collection of the content of the call, that is, a recording of the call. Given the packetization of the voice stream and its dispersal over an IP network, this is a technically difficult task. Some recent products can provide this capability; however, these products are very expensive.

At the time of this writing, a number of VoIP service providers have come up with solutions that take advantage of existing broadband availability in homes and businesses. VoIP service providers, such as Vonage and Packet8, offer their services online and sell analog telephone adapter kits in retail outlets. While they are not legally bound to do so, these services do offer some limited CALEA compliance via their softswitch, which can provide some details of the targeted call depending on the routing of the call. In short, market-leading softswitch vendors do have the capability to be CALEA compliant.

Architecture of WiMAX VoIP: Putting It All Together

What is the architecture for an alternative to the PSTN? The PSTN is comprised of three elements: access (the wires to a residence, for example), switching (the switches in the CO), and transport (long-distance ATM networks or IP fiber optic backbones). Figure 8-4 shows how voice services can be handled via an alternative network

Figure 8-4

An alternative to the PSTN: WiMAX as access, softswitch for switching, and IP backbones as transport

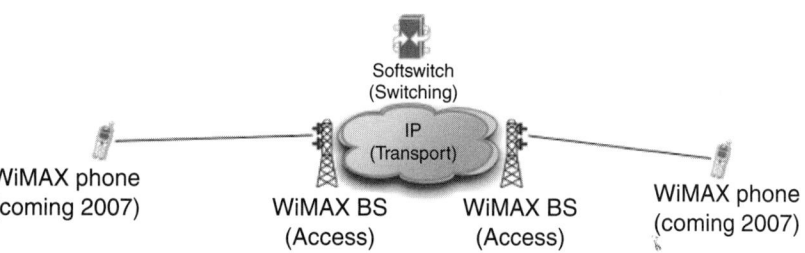

PSTN Bypass with WiMAX and VoIP

where access is performed by WiMAX (or associated protocols), switching is done by a softswitch, and transport is handled by IP fiber optic backbone. The common denominator in this alternative to the PSTN is VoIP. Wherever there is access to an IP stream, VoIP is possible. Softswitch technologies make managing voice traffic over an IP network possible.

WiMAX VoIP Phones

It won't be long until many WISPs begin to incorporate VoIP into their wireless service offerings. The following sections describe the mechanics of rolling out such services.

Case Study: AmberWaves WISP WiMAX VoIP   AmberWaves is a WISP in northwest Iowa. One of their clients has three offices with 35 employees linked by a WiMAX-like network. The greatest distance between the three offices is 19 miles.

This wireless network allows the firm to be its own internal data and voice service provider. The use of WiMAX-like service frees the firm from local and long-distance telephone bills. The end users report that the QoS on the network is better than with the frame relay circuit they used previously. Figure 8-5 shows how this could work with WiMAX.

Figure 8-5
Linking offices
with WiMAX
VoIP

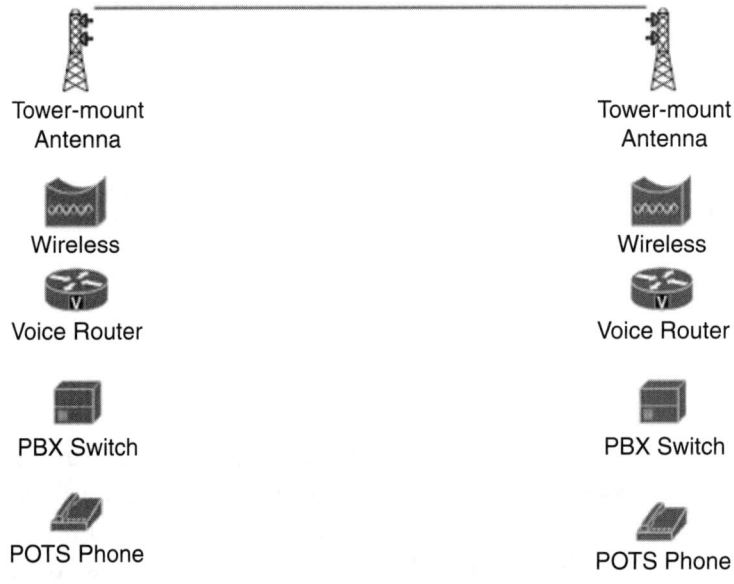

Conclusion

At the present time, the RBOCs of North America are losing thousands of lines per month; this is the first loss in coverage percentage-wise since the Great Depression. Most of the blame for these losses is placed on cell phones. SBC Communications reports a loss of three million phone lines (called access lines) between 2000 and 2002 and reports it will lose another three million lines in 2002. Other market analysts point to a number of influences.

Cell phones have claimed a number of those land lines. Many subscribers find cell phones more convenient and have "cut the wire." The monthly subscription cost of a cell phone has dropped, and many subscribers are dropping their land lines. Comparisons between the land line and cell phone seem to favor the cell phone's convenience over the land line's reputation for reliability and QoS.

Another explanation for one of the leading RBOC's loss of almost six million lines in a little over two years is broadband. Because the traditional 64 Kbps copper pair service provided to the majority of

North American residences and small businesses was designed entirely for voice service with a limited data service capability (56 Kbps with some 10 percent of households able to receive DSL service at data rates around 256 Kbps), there does not appear to be an appreciable level of "future proofing" built into the PSTN. Many DSL and cable television subscribers have either taken up VoIP applications for their voice service or have relied on cell phones for voice and their broadband connection for Internet access. DSL subscribers have canceled their second phone lines and now use their primary line for both DSL and telephone. Cable subscribers have canceled their land line altogether.

WiMAX IPTV

WISP WiMAX Triple Play?

Many telecommunications strategists speak of the elusive *triple play* of telecommunications where TV (video), voice, and data are all available from one service provider on one service medium and billed on one monthly, converged bill. At the current time, this is not entirely possible from existing service providers on existing infrastructure. Internet Protocol Television (IPTV) is a video technology that competes favorably with cable and broadcast television.

IPTV: Competing with Cable TV and Satellite TV

The rollout of digital networking infrastructure is opening the door for telcos and operators to offer converged services comprising broadband Internet access and IP-based TV and entertainment. TV (or video) over IP is a broad streaming solution that includes several applications, all of which can be implemented on WiMAX. IPTV is used in the following applications:

- TV (instead of cable TV) to the living room
- Time-shifted TV or personal video recorder (PVR)
- Interactive TV
- TV to the desktop

One of the inhibitors of broadband deployment in recent years has been the lack of broadband applications. This, in turn, resulted in low return on investment in broadband infrastructure. The viability of the broadband business model is becoming much more attractive with the introduction of IPTV services, which are a major revenue engine for telcos and ISPs.

With WiMAX IPTV, WISPs using WiMAX can offer a triple play of voice, video, and data to their subscribers. Customers receive converged services on a single pipe and interface with a single provider for all communication needs, resulting in easier technical maintenance, streamlined billing, and hence improved customer service. It is possible to target specific channels at small groups of viewers, based on predefined viewing profiles. Interactive IPTV also lets viewers create customized profiles themselves, based on their personal viewing habits. Table 9-1 provides a comparison of WiMAX to existing telecommunications infrastructures.

Table 9-1

WiMAX vs. Telco vs. Cable TV for Triple Play

Element	Legacy Telephone	Cable	WiMAX
Ability of infrastructure to deliver video, voice, and data	Largely voice-only; aging, non-video-capable, bandwidth-limited copper wire (mostly dial-up only); limited percentage of U.S. COs can deliver generic digital subscriber line (XDSL) service	Aging plant; mostly coax cable TV delivery only; increasing bidirectional cable modem service in cities and suburbs; most promising voice is third party (Vonage, Packet8, and so on)	High bandwidth capacity with least expensive infrastructure and operating costs to deliver voice, video, and data
Network management	Limited to CO and some outside plants; otherwise expensive truck roll	Limited network management system; truck rolls required	SNMP capable—can determine problems down to the IP address or device; truck roll for emergencies only
Bandwidth	XDSL for residential service usually less than 1 Mbps download; service limited to more densely populated markets	Focused on analog video; where available, most data services do not exceed 1 Mbps with limited QoS	Depending on BS, the "sky's the limit"; requires 2 Mbps for compressed MPEG 4 stream to deliver standard TV programming
Easily deployed to new markets?	With exception of fiber to the home at $1,400 per household, not economically feasible	Right-of-way and city franchise issues preclude timely deployment of service	No right-of-way issues; per FCC, no municipal concessions required; can be rapidly deployed

Regardless of the terminology, the process is the same: A TV program is converted to IP and streamed to the viewer. The same programming happens at the same time as cable or satellite TV. Assuming the viewer watches the programming on his or her TV set using an IP set top box (STP), the viewer's experience is no different than the experience of anyone else watching cable or satellite TV.

What's disruptive about this technology is that it's not limited to a traditional TV service provider. It is often called telco TV, as telephone companies feel the need to compete with cable TV companies offering voice services. In that scenario, the telephone company needs to offer broadband Internet access via variants of asymmetric digital subscriber line (ADSL).

Industry analysts refer to this cable TV vs. telco competition as a *duopoly*. WiMAX makes for a third force where WISPs offer data, voice, and the same TV programming as the other providers. A new term is creeping into that lexicon: WiMAX IPTV.

How It Works

Key technical components of the IPTV service provider's solution include a few key components that mesh traditional TV program distribution with IP technologies.

1. **Content and Programming**—The IPTV service provider has secured strategic transport agreements with national programmers and broadcasters (in North America, ESPN, CNN, and so on) to offer a competitive channel line-up, simplifying the acquisition process for the partner service provider (a WiMAX-powered ISP for example). The IPTV service provider receives this content directly from the programmers and feeds it into its encoding platform.

2. **Encoding**—The IPTV service provider encodes the video into MPEG-2 transport streams at a constant bit-rate, ensuring high-quality viewing while giving the service provider the confidence to guarantee two simultaneous streams per household over

WiMAX (requires a 10 Mbps downstream). The IPTV service provider's high-end encoding platform provides superior viewing quality.

3. **IP Streaming**—The MPEG-2 transport streams are encapsulated in UDP/IP and sent as individual multicast streams to the satellite UL. The IP streaming platform also applies IP QoS (via DiffServ code-point marking) and applies the IP multicast address for that channel.

4. **Satellite Transport**—The IPTV service provider then uplinks the IP multicast streams in a secure digital video broadcasting (DVB) format to the IPTV service provider's satellite. At the WISP's point-of-presence (POP), the IPTV service provider provides and installs the receive platform (including the satellite dish and receivers as well as decryption and demodulation equipment) required to convert the DVB format back to IP for handoff via redundant Gigabit Ethernet connections. The service provider can then deliver the video streams as native IP or encapsulate them in ATM.

5. **Local Encoding/Streaming (optional)**—The IPTV service provider also offers on-site encoding services to locally encode and stream local off-air (including Emergency Alert System) and regional and/or community programming. The IPTV service provider installs, configures, tests, and supports this service, and it is fully compatible with the IPTV service provider's national programming feeds.

6. **System Architecture**—Each component and stage of the IPTV service provider's network is fully redundant and proactively monitored and managed from the IPTV service provider network operations center (NOC). This level of management and reliability far exceeds that of its cable and direct broadcast satellite (DBS) competition.[1]

---

[1]"IPTV Transport Network," *Broadstream Communications*, available online at http://broadstream.com/prodserv/.

Bandwidth and Compression Technologies

IPTV requires bandwidths from around 3 Mbit/s minimum (depending on compression technology and desired resolution) in order to deliver broadcast quality video.[2] It is possible with reduced resolution to get acceptable picture quality down to 1.5 Mbit/s with standard MPEG-2 compression. Figure 9-1 compares IPTV to satellite and cable TV.

Other Video Revenue Streams

By the mid-1960s a majority of homes in the United States had TV. We would now call this *wireless residential video service*. Subscribers

Figure 9-1
IPTV infrastructure duplicates satellite and cable TV.

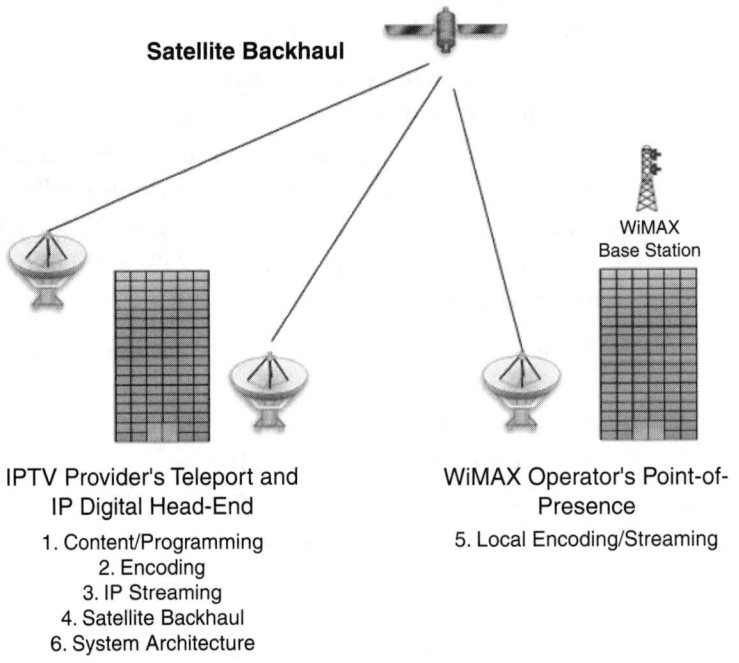

Satellite Backhaul

WiMAX
Base Station

IPTV Provider's Teleport and
IP Digital Head-End

1. Content/Programming
2. Encoding
3. IP Streaming
4. Satellite Backhaul
6. System Architecture

WiMAX Operator's Point-of-
Presence

5. Local Encoding/Streaming

[2]Helge Stephanson and Rolf Ollmar, "The Complete Guide to TVoIP," *Tandberg TV,* February 2, 2002, www.broadcastpapers.com/data/TandbergTVOverIP.pdf.

were limited in content to receiving three channels of programming focused on the evening hours known as prime time. Those subscribers were forced to be present in front of their video monitors at precisely the time of the broadcast. There was no means of storing the program for viewing at a later time. The coming of cable TV and videocassette recorders (VCRs) in the following decades added some flexibility to the TV viewing experience.

Before cable TV and VCRs, the subscribers were entirely at the mercy of the programmers. They had to watch what the programmers offered. The ability to choose programming drove the growth of cable TV and VCRs, leading to a myriad of new businesses including cable TV companies and video rental firms. The production and distribution costs were very high for most programming (films and prime time TV shows). This presented a high barrier to entry for any competitors.

Video on Demand

In October of 2002, a startup firm named Cflix launched a paid video download service offering a variety of feature length films and some video serials, such as the popular animation "South Park." One month later, a consortium of Hollywood firms launched a service called MovieLink, which offers recent Hollywood releases for a fee per download via broadband Internet connections. Starz Ticket offers newly released films for a monthly subscription of $12.95 (in the United States only at the present time).

A Cflix subscription, which includes some basic programming, costs $4 per month—considerably less expensive than other video services. Cflix subscribers pay an additional $1.99 for older movies and $3.99 for new releases. They can attach equipment to their computers that allows them to watch the movies on a TV set.[3]

Downloading movies via streaming video is not new. File sharing of video files, including feature length films, has been available

---

[3]Dan Luzadder, "Video Service Gives It That New College Try," *Denver Post,* October 22, 2002, www.denverpost.com/cda/article/print/0,1674,36%257E33%257E940472,00.html.

online for years. What is new is the commercialization of this practice made possible by residential broadband. The deployment of WiMAX will accelerate this trend.

Personal Video Recorder

Due simply to the broadband connection made possible by WiMAX, personal video recorders (PVRs) will grow in popularity. The PVR, also called digital video recorder (DVR), is a consumer electronics device that records television shows to a hard disk in digital format. This makes the *time shifting* feature (traditionally done by a VCR) much more convenient. It also allows for "trick modes," such as pausing live TV, instantly replaying interesting scenes, and skipping advertising. Most PVR recorders use the MPEG format for encoding analog video signals.

Many satellite and cable companies are incorporating PVR functions into their STB, such as with DirecTiVo. In this case, encoding in the PVR is not necessary, as the satellite signal is already a digitally encoded MPEG stream. The PVR simply stores the digital stream directly to disk.

Conclusion: A TV Station Called WiMAX

By offering the same programming at the same time on the same "channels" as cable TV or satellite TV, WISPs will give prospective subscribers reason to buy their service. A residential customer may not find WiMAX as a broadband solution compelling in and of itself. They may only see it as "faster e-mail." However, couple that service with VoIP, and the subscriber sees a value in subscribing to a broadband service such as WiMAX. Go one step further and offer video services competitive with their existing cable TV or satellite service, and the sale is a done deal.

Regulatory Aspects of WiMAX

For many service providers, the gating factor in making the decision to deploy a WiMAX network may revolve around actions of regulatory agencies. The first concern is projected cost structures in using licensed vs. unlicensed spectrum. In an ideal environment, spectrum is free, and there is no interference from other broadcasters. Even when this is the case, it may not always be so. This chapter will first outline what the operator needs to know regarding unlicensed frequencies and then will cover the FCC's move to liberalize spectrum policy (that is, make more of it available to operators, especially in light of the FCC's initiative to boost access to broadband to Americans by whatever means).

Operate Licensed or Unlicensed?

An objection often raised about WiMAX applications is that because some spectrum potentially used by WiMAX is unlicensed, it will inevitably become overused (like common land in the "tragedy of the commons") so that it becomes unusable. At this time, the government (United States or other) will step in to control the spectrum, making it "not free" and thus costing the service provider his or her profit margin and relegating the market to deep-pocketed monopolists.

In this scenario, the service provider can buy rights to a licensed frequency either directly from the FCC, from another operator, or through a frequency broker. How much does this cost? That depends on the location (urban, suburban, or rural) and the frequency. (Other operators may highly desire this frequency, or it may appear to have no market value and be priced accordingly.) Once that operator's claim to that frequency is formalized, he or she is protected from interference by other operators.

This chapter will explore first the considerations wireless service providers should take into account when deploying service on unlicensed WiMAX bands. Next, the chapter will explore a new initiative from the FCC, which heralds a change in spectrum management and which may actually serve to liberalize the FCC's approach to what spectrum is unlicensed.

Table 10-1

Considerations
in What
Spectrum to
Use

Band	5.8 GHz	2.5 GHz	3.5 GHz
Licensing	License-exempt worldwide	Licensed in U.S., Canada, some of Latin America	Unlicensed in Europe, Latin America, Asia
Cost	N/A	Varies, can also lease from license holder	Varies, can also lease from license holder
Spectrum	Up to 125 MHz in U.S.	22.5 MHz/license in U.S.	Varies by country
Allowable transmit power	U.S.: Max power to antenna 1 watt, Max EIRP +53 dBm (200 watts)	U.S.: Max EIRP +55 dBW	Per ETSI: 3 watts (+35dBm) max to antenna (varies by country)
Interference control	Restrict deployment to less than $1/2$ the available spectrum, use auto channel select and coordinate between operators	Protected by license assignment; no two operators assigned same frequency in same area	Protected by license assignment; no two operators assigned same frequency in same area
BSs required	Higher BS capacity results in fewer BS sites to achieve area coverage	More BS sites to meet capacity requirements due to limited spectrum assignment	More BS sites to meet capacity requirements due to limited spectrum assignment
Indoor and outdoor customer premise equipment (CPEs)	Can support indoor CPE at customer site within 800 meters from BS; outdoor CPEs must be deployed elsewhere; RESULT: Higher average CPE cost and higher average installation cost	Supports a high percentage of indoor CPEs in capacity limited deployments; RESULT: Lower average CPE cost and lower average installation cost	Supports a high percentage of indoor CPEs in capacity limited deployments; RESULT: Lower average CPE cost and lower average installation cost

Finally the chapter will cover an initiative in the U.S. Congress to free more spectrum for use as broadband wireless Internet applications. If anything, it appears that the United States government is developing a policy to encourage the use of unlicensed spectrum.

Current Regulatory Environment

Even though WiMAX can operate in unlicensed spectrum, a service provider must know a number of things in order to stay out of trouble with state and federal authorities. The following pages will outline the most prominent problem areas.

Spectrum is managed by a number of different organizations. The most visible to the general public is the FCC. The FCC manages civilian, state, and local government usage of the radio spectrum. The FCC regulations are contained in the "Code of Federal Regulations, Title 47."

At the present time, the FCC has very limited resources for enforcement, as the trend for the last couple of decades is deregulating and reducing staffing in the enforcement bureaus. Also, the National Telecommunications and Information Administration (NTIA) works with the Interdepartmental Radio Advisory Committee (IRAC), which manages federal use of the spectrum.

The following pages offer a brief overview of what a service provider needs to be concerned about when operating in unlicensed spectrum. Tim Pozar of the Bay Area Wireless Users Group provided this synopsis, based on many years of experience advising friends and clients as to what they can and cannot do with unlicensed spectrum. The treatise was originally intended to provide guidelines for 802.11 operators, but the law applies equally to 802.16 operators.

Power Limits

Although WiMAX can do 70 Mbps over 30 miles, it must comply with the power restrictions for that band if it is to operate in an unlicensed frequency. Ideally, a well-engineered path will have just the amount of power required to get from point A to point B with good reliability. Good engineering will limit the signal to the area being served, which has the effect of reducing interference and providing a more efficient use of the spectrum. Using too much power will cover more area than is needed and can potentially interfere with other users of the band.

WiMAX 802.16—Its Relationship to FCC Part 15, Section 247

If service providers intend to use unlicensed spectrum with their WiMAX deployment, it would be a good idea to have a thorough understanding of FCC Part 15.

Point-to-Multipoint WiMAX service providers who wish to operate under this section are allowed up to 30 dBm or 1 watt of Transmitter Power Output (TPO) with a 6 dBi antenna or 36 dBm or 4 watts effective radiated power over an equivocally isotropic radiated power (EIRP) antenna. The TPO needs to be reduced 1 dB for every dB of antenna gain over 6 dBi.

Point-to-Point The FCC encourages directional antennas to minimize interference to other users. The FCC, in fact, is more lenient with point-to-point links by requiring only the TPO to be reduced by $1/3$ of a dB instead of a full dB for point-to-multipoint. More specifically, for every 3 dB of antenna gain over a 6 dBi antenna, a WISP needs to reduce the TPO 1 dB below 1 watt. For example, a 24 dBi antenna is 18 dB over a 6 dBi antenna. This requires lowering a 1 watt (30 dBm) transmitter $18/3$ or 6 dB to 24 dBm or .25 watt.

802.16—FCC Part 15, Section 407

So what part of Part 15 applies to WiMAX operations in the 5 GHz range? The following paragraphs will outline the law for this spectrum.

Point-to-Multipoint   As described earlier, the U-NII band is chopped into three sections. The "low" band runs from 5.15 GHz to 5.25 GHz and has a maximum power of 50 mW (TPO). This band is meant to be in-building only, as defined by the FCC's Rules and Regulations (R&R) Part 15.407 (d) and (e):

(d) Any U-NII device that operates in the 5.15–5.25 GHz band shall use a transmitting antenna that is an integral part of the device.

(e) Within the 5.15–5.25 GHz band, U-NII devices will be restricted to indoor operations to reduce any potential for harmful interference to co-channel MSS operations.[1]

The "middle" band runs from 5.25 GHz to 5.35 GHz, with a maximum power limit of 250 mW. Finally, the "high" band runs from 5.725 GHz to 5.825 GHz, with a maximum transmitter power of 1 watt and antenna gain of 6 dBi or 36 dBm or 4 watts EIRP.

Point-to-Point   The FCC does give some latitude to point-to-point links in 15.407(a)(3). For the 5.725 GHz to 5.825 GHz band, the FCC allows a TPO of 1 watt and up to a 23 dBi gain antenna without reducing the TPO 1 dB for every 1 dB of gain over 23 dBi.

15.247(b)(3)(ii) does allow the use of any gain antenna for point-to-point operations without having to reduce the TPO for the 5.725 GHz to 5.825 GHz band.

Interference

The raison d'être of the Radio Act of 1927 was to ensure that radio operators could operate with minimum interference from other broadcasters. Part 15 was established to provide a framework for those operating in the unlicensed spectrum to avoid interfering with each other.

Description   Of course, interference is typically the state of the signal one is interested in while it's being destructively overpowered

---

[1]Tim Pozar, "Regulations Affecting 802.11," June 6, 2002, www.lns.com/papers/ part15/.

by a signal one is not interested in. The FCC has a specific definition of *harmful interference*:

Part 15.3(m) Harmful interference.

Any emission, radiation or induction that endangers the functioning of a radio navigation service or of other safety services or seriously degrades, obstructs or repeatedly interrupts a radio communications service operating in accordance with this chapter.

As there may be other users of this band, interference will be a factor in WiMAX deployments. The 2.4 GHz band is often more congested than the 5.8 GHz band, but both have their co-users. The following subsections will describe the other users of this spectrum and what interference mitigation may be possible for each.

Devices that Fall into Part 15 (2400–2483 MHz) Table 10-2 lists which FCC regulations apply to which frequency bands. Table 10-3 lists the spectrum bands of ISM. Unlicensed telecommunications devices like cordless phones, home spy cameras, and Frequency Hopping (FHSS) and Direct Sequence (DSSS) Spread Spectrum LAN transceivers fall into Part 15 (2400–2483 MHz). Operators have no priority over or parity with any of these users. Any device that falls into Part 15 must not cause harmful interference to and must accept interference from all licensed and all legally operating Part 15 users.

Operators of other licensed and nonlicensed devices can inform users of interference and require that they terminate operation. This source needn't be a Commission representative.

Part 15.5(b) operation of an intentional, unintentional, or incidental radiator is subject to the conditions that no harmful interference is caused and that interference must be accepted that may be caused by the operation of an authorized radio station, by another intentional or unintentional radiator, by industrial, scientific and medical (ISM) equipment, or by an incidental radiator (or basically everything).

Table 10-2

Spectrum
Allocation for
U-NII and
Co-Users

Part/Use	Start GHz	End GHz
Part 87	0.4700	10.5000
Part 97	2.3900	2.4500
Part 15	2.4000	2.4830
Fusion Lighting	2.4000	2.4835
Part 18	2.4000	2.5000
Part 80	2.4000	9.6000
ISM—802.11b	2.4010	2.4730
Part 74	2.4500	2.4835
Part 101	2.4500	2.5000
Part 90	2.4500	2.8350
Part 25	5.0910	5.2500
U-NII Low	5.1500	5.2500
U-NII Middle	5.2500	5.3500
Part 97	5.6800	5.9250
U-NII High	5.7250	5.8250
ISM	5.7250	5.8500
Part 18	5.7250	5.8250

Source: Tim Pozar Bay Area Wireless Users Group from FCC sources

15.5(c) The operator of a radio frequency device shall be required to cease operating the device upon notification by a commission representative that the device is causing harmful interference. Operation shall not resume until the condition causing the harmful interference has been corrected.

Table 10-3

United States
ISM Channel
Allocations

Channel	Bottom (GHz)	Center (GHz)	Top (GHz)
1	2.401	2.412	2.423
2	2.406	2.417	2.428
3	2.411	2.422	2.433
4	2.416	2.427	2.438
5	2.421	2.432	2.443
6	2.426	2.437	2.448
7	2.431	2.442	2.453
8	2.436	2.447	2.458
9	2.441	2.452	2.463
10	2.446	2.457	2.468
11	2.451	2.462	2.473

Source: Tim Pozar Bay Area Wireless Users Group from FCC sources

Devices That Fall into the U-NII Band Unlike the 2.4 GHz band, this band does not have overlapping channels. The lower U-NII band has eight 20 MHz wide channels. One can use any of the channels without interfering with other radios on other channels that are within "earshot." Ideally, it would be good to know what other Part 15 users are out there.

Industrial, Scientific, and Medical (ISM)—Part 18 This is also an unlicensed service. Typical ISM applications are the production of physical, biological, or chemical effects such as heating, ionization of gases, mechanical vibrations, hair removal, and acceleration of charged particles. Users are ultrasonic devices, such as jewelry cleaners, ultrasonic humidifiers, and microwave ovens; medical devices, such as diathermy equipment and magnetic resonance imaging equipment (MRI); and industrial devices, such as paint dryers (18.107). RF should be contained within the devices, but other users must accept interference from these devices. Part 18 frequencies that could affect WiMAX devices are 2.400 to 2.500 GHz and 5.725 GHz to 5.875 GHz. As Part 18 devices are unlicensed and operators are likely clueless on the impact, it will be difficult to coordinate with them. Part 18 also covers fusion lighting.

Satellite Communications—Part 25 This part of the FCC's rules is used for the UL or DL of data, video, and so on to/from satellites in Earth's orbit. One band that overlaps the U-NII band is reserved for Earth-to-space communications at 5.091 to 5.25 GHz. Within this spectrum 5.091 to 5.150 GHz is also allocated to the fixed-satellite service (Earth-to-space) for nongeostationary satellites on a primary basis. The FCC is trying to decommission this band for "feeder" use to satellites, as "after 01 January 2010, the fixed-satellite service will become secondary to the aeronautical radionavigation service." A note in Part 2.106 [S5.446] also allocates 5.150 to 5.216 GHz for a similar use, except it is for space-to-Earth

Table 10-4

Popular Unlicensed Spectra and Their Associated Power Data

Frequency Range (MHz)	Bandwidth (MHz)	Max Power at Antenna	Max EIRP	Notes
2,400–2,483.5	83.5	1 W (+30dBm)	4 W (+36dBm)	Point-to-point
		1 W (+30dBm)		Point-to-multipoint following 3:1 rule
5,150–5,250	100	50 Mw	200 mW (+23dBm)	Indoor use; must have integral antenna
5,250–5,350	100	250 mW (+24dBm)	1 W (+30dBm)	
5,725–5,825	100	1 W (+30dBm)	200 W (+53dBm)	

communications. There is a higher chance of interfering with these installations, as Earth stations are dealing with very low signal levels from distance satellites.

Broadcast Auxiliary—Part 74 Normally the traffic is electronic news gathering (ENG) video links going back to studios or television transmitters. These remote vehicles, such as helicopters and trucks, need to be licensed. Only Part 74 eligibles, such as TV stations, networks, and so on, can hold these licenses (74.600). Typically these transmitters are scattered all around an area, as TV remote trucks can go anywhere. This can cause interference to WiMAX gear, such as BS deployed with omnidirectional antennas servicing an area. Also the receive points for ENG are often mountaintops and towers. Depending on how WiMAX BSs are deployed at these same locations, they could cause interference to these links. Wireless providers should consider contacting a local frequency coordinator for Part 74 frequencies that would be affected. ENG frequencies that overlap ISM devices are 2.450 to 2.467 GHz (channel A08) and 2.467 to 2.4835 GHz (channel A09), (Part 74.602).

Land Mobile Radio Services—Part 90 For subpart C of this part, a user can be anyone engaged in a commercial activity. They can use from 2.450 to 2.835 GHz but can only license 2.450 to 2.483 GHz. Users in subpart B would be local governments, including organizations such as law enforcement agencies, fire departments, and so on. Some uses may be video DLs for flying platforms such as helicopters, aka terrestrial surveillance. Depending on the commercial or government agency, coordination goes through different groups like Association of Public Safety Communications Officials (APCO). Consider going to their conferences. Also, try to network with engineering companies that the government outsources to for their frequency coordination.

Amateur Radio—Part 97 Amateur radio frequencies that overlap ISM are 2.390 to 2.450 GHz and 5.650 to 5.925 GHz for U-NII. They are primary from 2.402 to 2.417 GHz and secondary at 2.400 to 2.402 GHz. There is a Notice of Proposed Rule Making (NPRM) with the FCC to change the 2.400 to 2.402 GHz to primary.

Fixed Microwave Services—Part 101   Users are known as local television transmission service (LTTS) and private operational fixed point-to-point microwave service (POFS). This band is used to transport video. Users are allocated from 2.450 to 2.500 GHz.

Federal Usage (NTIA/IRAC)   The federal government uses this band for radiolocation or radionavigation. Several warnings in the FCC's Rules and Regulations disclose this fact. In the case of 802.16b, a note in the rules warns:

> 15.247(h) Spread spectrum systems are sharing these bands on a noninterference basis with systems supporting critical government requirements that have been allocated the usage of these bands, secondary only to ISM equipment operated under the provisions of Part 18 of this chapter. Many of these government systems are airborne radiolocation systems that emit a high EIRP, which can cause interference to other users.

In the case of U-NII, the FCC has a note in Part 15.407 stating that:

> Commission strongly recommends that parties employing U-NII devices to provide critical communications services should determine if there are any nearby government radar systems that could affect their operation.

Laws on Antennas and Towers

Many a local zoning board have found telecommunications towers to be considered "unsightly." How is an operator to deal with such allegations?

FCC Preemption of Local Law   The installation of antennas may run counter to local ordinances and homeowner agreements that would prevent installations. Thanks to the Satellite Broadcasting and Communications Association (SBCA), who lobbied the FCC, the FCC has stepped in and overruled these ordinances and agreements.

This ruling from the FCC should only apply to broadcast signals such as TV, DBS, or MMDS. It could be argued that the provision for MMDS could cover wireless data deployment.

Height Limitations   The placement of towers and other broadcast-related equipment could spark any series of "federal cases" or other lengthy disputes regarding which government has what jurisdiction on broadcasting equipment.

Local Ordinances   Most if not all cities regulate the construction of towers. There are maximum height zoning regulations regarding the antenna/tower (residential or commercial), construction, and aesthetics (for example, what color, how hidden).

FAA and the FCC Tower Registration   The FAA is very concerned about objects that airplanes might bump into. Part 17.7(a) of the FCC R&R describes "any construction or alteration of more than 60.96 meters (200 feet) in height above ground level at its site."[2]

New Unlicensed Frequencies

In June 2004, the FCC had recently approved plans to improve the management of a block of radio spectrum, 2.495 GHz to 2.690 GHz, to ease the way for the wider adoption of wireless broadband access.[3] Working its way through the U.S. Congress is the Jumpstart Broadband Bill (aka Boxer-Allen Bill—U.S. Senate), which would add 255 MHz in the 5 GHz unlicensed band. The bill is part of a wider move to bolster wireless broadband as a "third leg" to the broadband stool of cable and DSL (cable TV and telephone companies).[4]

---

[2]Tim Pozar, "Regulations Affecting 802.16 Deployment," white paper from Bay Area Wireless Users Group, pp. 2–7, 10–11, www.lns.com/papers/part15/.

[3]Richard Shim, "FCC Cleans Up Spectrum for Wireless Broadband," June 10, 2004, *CNET*, http://news.com.com/FCC+cleans+up+spectrum+for+wireless+broadband/2100-1034_3-5230766.html?tag=st.rc.targ_mb.

[4]Roy Mark, "Senators Aim to Wirelessly Jumpstart Broadband," November 20, 2002, http://siliconvalley.internet.com/news/article.php/1545891.

In March 2005, the FCC issued an order to open the 3.650–3.7 GHz spectrum for wireless broadband services. The licensing scheme that the FCC adopts for this band will provide an opportunity for the introduction of a variety of new wireless broadband services and technologies, such as WiMAX. Additionally, the actions the FCC takes herein for the 3,650 MHz band will allow further deployment of advanced telecommunications services and technologies to all Americans, especially in the rural heartland, thus promoting the objectives of Section 706 of the Telecommunications Act of 1996.[5] The chief caveat of this order is that these transmissions cannot occur near satellite ground stations listed in the order.

Unlicensed Frequencies Summary

Although frequencies in the ISM and U-NII bands are unlicensed (that is, free), they are not without restrictions. Power restrictions may limit the potential for a WiMAX operator to project the full potential of the platform to transmit a given bandwidth over a given distance. Operators considering using WiMAX on unlicensed spectrum should perform a thorough site survey to determine potential conflicts with the law and fellow broadcasters for the given location, frequency, and power level on which they intend to operate.

The FCC New Spectrum Policy

The American spectrum management regime is approximately 90 years old. In the opinion of former FCC chairman Michael Powell, it needs a hard look and a new direction. Historically, spectrum policy has four underlying core assumptions: (1) unregulated radio interference will lead to chaos; (2) spectrum is scarce; (3) government command and control of the scarce spectrum resource is the only way chaos can be avoided; and (4) the public interest centers on government choosing the highest and best use of the spectrum.

---

[5]Federal Communications Commission, "Report and Order of Opinion and Order 05-56," March 16, 2005, p. 2.

Four Problem Areas in Spectrum Management and Their Solutions

There are four problem areas the FCC will have to work its way through in order to liberalize spectrum policy. The payoff is a more functional spectrum policy to meet the needs of telecommunications users.

Interference—The Problem From 1927 until today, interference protection has always been at the core of federal regulators' spectrum mission. The Radio Act of 1927 empowered the Federal Radio Commission to address interference concerns. Although interference protection remains essential to our mission, interference rules that are too strict limit users' ability to offer new services; whereas rules that are too lax may harm existing services. I believe the Commission should continuously examine whether there are market or technological solutions that can—in the long run—replace or supplement pure regulatory solutions to interference.

The FCC's current interference rules were typically developed based on the expected nature of a single service's technical characteristics in a given band. The rules for most services include limits on power and emissions from transmitters. Each time the old service needs to evolve with the demands of its users, the licensee has to come back to the Commission for relief from the original rules. This process is not only inefficient; it can stymie innovation.

Due to the complexity of interference issues and the RF environment, interference protection solutions may be largely technology driven. Interference is not solely caused by transmitters, which many seem to assume, and on which our regulations are almost exclusively based. Instead, interference is often more a product of receivers; that is, receivers are too dumb or too sensitive or too cheap to filter out unwanted signals. Yet, the FCC's decades-old rules have generally ignored receivers. Emerging communications technologies are becoming more tolerant of interference through sensory and adaptive capabilities in receivers. That is, receivers can "sense" what type of noise or interference or other signals are operating on a given channel and then "adapt," so that they transmit on a clear channel that allows them to be heard.

Both the complexity of the interference task—and the remarkable ability of technology (rather than regulation) to respond to it—are most clearly demonstrated by the recent success of unlicensed operations. According to the Consumer Electronics Association, a complex variety of unlicensed devices—including garage and car door openers, baby monitors, family radios, wireless headphones, and millions of wireless Internet access devices using Wi-Fi technologies—is already in common use. Yet despite the sheer volume of devices and their disparate uses, manufacturers have developed technology that allows receivers to sift through the noise to find the desired signal.

Interference—The Solution   Legal approaches to interference mitigation may often be the easier solution. Legislation sorely needs to be updated to accommodate for advances in technology.

Interference Protection   The Interference Protection Working Group (Working Group) of the FCC's Spectrum Policy Task Force recommended that the FCC should consider using the Interference Temperature metric as a means of quantifying and managing interference. As introduced in this report, *interference temperature* is a measure of the RF power available at a receiving antenna to be delivered to a receiver—power generated by other emitters and noise sources. More specifically, it is the temperature equivalent of the RF power available at a receiving antenna per unit bandwidth, measured in units of degrees Kelvin. As conceptualized by the Working Group, the terms "interference temperature" and "antenna temperature" are synonymous. The term "interference temperature" is more descriptive for interference management. For a technical description of interference, see Chapter 6.

Like other representations of radio signals, instantaneous values of interference temperature would vary with time and, thus, would need to be treated statistically. The Working Group envisions that interference "thermometers" could continuously monitor particular frequency bands, measure and record interference temperature values, and compute appropriate aggregate value(s). These real-time values could govern the operation of nearby RF emitters. Measurement devices could be designed with the option to include or exclude the on-channel energy contributions of particular signals with known characteristics such as the emissions of users in geographic areas and bands where spectrum is assigned to licensees for exclusive use.

The FCC could use the interference temperature metric to set maximum acceptable levels of interference, thus establishing a worst case environment in which a receiver would operate. Interference temperature thresholds could thus be used, where appropriate, to define interference protection rights.

The time has come to consider an entirely new paradigm for interference protection. A more forward-looking approach requires that there be a clear quantitative application of what is acceptable interference for both license holders and the devices that can cause interference. Transmitters would be required to *ensure* that the interference level—or interference temperature—is not exceeded. Receivers would be required to *tolerate* an interference level.

Rather than simply saying a transmitter cannot exceed a certain power, the industry instead would utilize receiver standards and new technologies to ensure that communication occurs without interference and that the spectrum resource is fully utilized. So, for example, perhaps services in rural areas could utilize higher power levels because the adjacent bands are less congested, therefore decreasing the need for interference protection.[6]

From a simplistic and physical standpoint, any transmission facility requires a transmitter, a medium for transmission, and a receiver. Focus on receiver characteristics has not been high in past spectrum-use concerns; hence, a shift in focus is in order. The Working Group believes that receiver reception factors, including sensitivity, selectivity, and interference tolerance, need to play a prominent role in spectrum policy.[7]

Spectrum Scarcity—The Problem Much of the Commission's spectrum policy was driven by the assumption that there is never enough for those who want it. Under this view, spectrum is so scarce

---

[6]Michael Powell, "Broadband Migration—New Directions in Wireless Policy," speech to Silicon Flatirons Conference, University of Colorado, Boulder, October 30, 2002.

[7]Federal Communications Commission Spectrum Policy Task Force, "Report of the Interference Protection Working Group," November 15, 2002, p. 25.

that government, rather than market forces, must determine who gets to use the spectrum and for what. The spectrum scarcity argument shaped the Supreme Court's *Red Lion* decision, which gave the Commission broad discretion to regulate broadcast media on the premise that spectrum is a unique and scarce resource. Indeed most assumptions that underlie the current spectrum model derive from traditional radio broadcasting and are oblivious to wireless broadband Internet applications.

The Commission has recently conducted a series of tests to assess actual spectrum congestion in certain locales. These tests, which were conducted by the Commission's Enforcement Bureau in cooperation with the Task Force, measured use of the spectrum at five major cities in the United States. The results showed that although some bands were heavily used, others either were not used or were used only part of the time. It appeared that these "holes" in bandwidth or time could be used to provide significant increases in communication capacity without impacting current users through use of new technologies. These results call into question the traditional assumptions about congestion. Indeed, most of spectrum is apparently not in use most of the time.

Today's digital migration means that more and more data can be transmitted in less and less bandwidth. Not only is less bandwidth used, but innovative technologies, like software-defined radio and adaptive transmitters, can bring additional spectrum into the pool of spectrum available for use.

Spectrum Scarcity—The Solution   In analyzing the current use of spectrum, the Task Force took a unique approach: For the first time, they looked at the *entire* spectrum, not just one band at a time. This review prompted a major insight—there is a substantial amount of *white space* out there that is not being used by anybody. The ramifications of this insight are significant. Although spectrum *scarcity* is a problem in some bands some of the time, the larger problem is spectrum *access:* or how to get to and use those many areas of the spectrum that are either underutilized or not used at all.

One way the Commission can take advantage of this white space is by facilitating access in the time dimension. Since the beginning of spectrum policy, the government has parceled this resource in frequency and in space. The FCC historically permitted use in a particular band over a particular geographic region, often with an expectation of perpetual use. The FCC should also look at *time* as an additional dimension for spectrum policy. How well could society use this resource if FCC policies fostered access in frequency, space, and time?

Technology has facilitated access to spectrum in the time dimension, which will lead to more efficient use of the spectrum resource. For example, a software-defined radio may allow licensees to dynamically "rent" certain spectrum bands when they are not in use by other licensees. Perhaps a mobile wireless service provider with software-defined phones will lease a local business's channels during the hours the business is closed. Similarly sensory and adaptive devices may be able to "find" spectrum open space and utilize it until the licensee needs those rights for its own use. In a commercial context, secondary markets can provide a mechanism for licensees to create and provide opportunities for new services in distinct slices of time. By adding another meaningful dimension, spectrum policy can move closer to facilitating consistent availability of spectrum and further diminish the scarcity rationale for intrusive government action.

Government Command and Control—The Problem   The theory back in the 1930s was that only government could be trusted to manage this scarce resource and ensure that no one got too much of it. Unfortunately, spectrum policy is still predominantly a *command and control* process that requires government officials—instead of spectrum users—to determine the best use for spectrum and make value judgments about proposed, and often overhyped, uses and technologies. It is an entirely reactive and too easily politicized process.

In the last 20 years, two alternative, very flexible models to command and control the spectrum have developed. The first is the *exclusive use* or quasi-property rights model. This model provides exclusive, licensed rights to flexible-use frequencies, subject only to limitations on harmful interference. These rights are freely transferable. The second is the *commons* or *open access* model. This model allows users to share frequencies on an unlicensed basis with usage rights that are governed by technical standards but without any right to protection from interference. The Commission has employed both models with significant success. Licensees in mobile wireless services have enjoyed quasi-property right interests in their licensees and transformed the communications landscape as a result. In contrast, the unlicensed bands employ a commons model and have enjoyed tremendous success as hotbeds of innovation.

Government Command and Control of the Spectrum—The Solution   Historically the Commission often limited flexibility via command and control regulatory restrictions on which services licensees could provide and who could provide them. Any spectrum users that wanted to change the power of their transmitter, the nature of their service, or the size of an antenna had to come to the Commission to ask for permission, wait the corresponding period of time, and only then, if relief was granted, modify the service. Today's marketplace demands that the FCC provide license holders with greater flexibility to respond to consumer wants, market realities, and national needs without first having to ask for the FCC's permission. License holders should be granted the maximum flexibility to use—or allow others to use—the spectrum, within technical constraints, to provide any services demanded by the public. With this flexibility, service providers can be expected to move spectrum quickly to its highest and best use.

Public Interest—The Problem   The fourth and final element of traditional spectrum policy is the *public interest* standard. The phrase (or something similar) "public interest, convenience, or necessity" was a part of the Radio Act of 1927 and likely came from other utility regulation statutes. The standard was largely a response to the interference and scarcity concerns that were created in the absence of such a discretionary standard in the 1912 Act. This "public interest, convenience, and necessity" became a standard by which to judge between competing applicants for a scarce resource—and a tool for ensuring interference did not occur. The public interest under the command and control model often decided which companies or government entities would have access to the spectrum resource. At that time, spectrum was not largely a consumer resource but, rather, was accessed by a relatively select few. However, Congress wisely did not create a static public interest standard for spectrum allocation and management.

Serving the Public Interest in Spectrum Policy—The Solution The FCC should develop policies that avoid interference rules that are barriers to entry, that assume a particular proponent's business model or technology, and that take the place of marketplace or technical solutions. Such a policy must embody what we have seen benefit the public in every other area of consumer goods and services: choice through competition and limited but necessary government intervention into the marketplace to protect such interests as access to people with disabilities, public health, safety, and welfare.

Recent Statements from the FCC on Broadband and Spectrum Policy

A recurring objection to WiMAX is pessimism toward what role regulators will take: "Won't 'they' take away free spectrum and prop up the monopolistic incumbents?" Indications from the FCC seem to point in the opposite direction. Below are recent comments by former FCC chair Michael Powell. Readers are invited to make their own conclusions as to whether this will be enough remedy in time to launch the WiMAX revolution.

> Earlier this year [2004], I [FCC chair Michael Powell] created the Wireless Broadband Access Task Force to review our wireless broadband policies and to identify areas where additional Commission action, or restraint, could facilitate further deployment. The task force has identified several key issues in this regard.
>
> *First, we need more broadband spectrum.* In this era of increasingly intensive spectrum use, we must continue to strive to provide opportunities for new and enhanced spectrum-based services. I applaud the Administration's decision to undertake a comprehensive review of spectrum policy. The reports of the President's Spectrum Policy Initiative offer much food for

thought about these timely issues. The significant spectrum reforms that we at the FCC have worked so hard to identify and implement over the last two years, coupled with the results of the President's Spectrum Policy Initiative, will help enable us to craft policies that will facilitate delivery of wireless broadband services to the American people.

The FCC is moving aggressively to put valuable spectrum on the market through auctions. In January, the Commission will auction over 200 broadband PCS C and F block licenses. In addition, we are working collaboratively with our colleagues at NTIA to move forward expeditiously to an auction of spectrum at 2 GHz for advanced wireless services. We also greatly appreciate Congress' efforts to craft the Spectrum Relocation Trust Fund to ensure that the relocation of military operations that currently use this spectrum can be adequately funded with the proceeds of this auction. I urge Congress to pass this legislation as quickly as possible.

A second key conclusion is that we need greater access to the spectrum that is in the market. One significant finding of our task force effort was that most of the spectrum is not being used most of the time. This means that rather than scarcity being the problem, the real problem is how to get access to spectrum. We believe technology is going to usher in the possibility of much more dynamic use of frequencies without unacceptable interference.

At the federal level, we must push for procompetitive, market-based policies for all broadband technologies in order to allow the various platforms to compete freely and fully. Wireless, cable, DSL, satellite, and power lines should compete where it makes sense for them to compete and become integrated where they are complementary. In such a market, consumers benefit greatly, as the market itself can change to meet consumers' needs far faster than regulators could act to address consumers' concerns.[8]

---

[8]Michael K. Powell, "The Wireless Broadband Express" (remarks, CTIA Wireless I.T. & Entertainment Convention, San Francisco, October 26, 2004).

Our unlicensed rules have been a hotbed for wireless broadband innovation—spawning new industries like your own and encouraging significant capital investment. It is estimated that by next year, sales of wireless networking equipment will exceed $5 billion. Our regulatory flexibility in this area has helped to enable this thriving industry.

We continue to look for more ways to encourage growth of unlicensed wireless broadband services. Last year, we made an additional 255 MHz of spectrum available in the 5 GHz region of the spectrum—adding a sizable chunk of spectrum to that already available for unlicensed devices. We also made spectrum available in the upper reaches of the spectrum—above 70 GHz—on an unlicensed and very lightly licensed basis. Technologies that use this new spectrum frontier are rapidly maturing and new services are on the horizon. We are also in the process of considering additional spectrum bands for use by unlicensed devices—the so-called spectrum "white spaces" between the channels assigned for TV broadcast services and 50 MHz of spectrum in the 3,650 MHz band.[9]

Conclusion

This chapter outlines the current regulatory regime for WiMAX operators. The chapter answers the objection that there is too little spectrum available for a mass-market deployment of WiMAX. Recent studies and pronouncements by the FCC and members of the U.S. Senate indicate support for reforming the spectrum policy in promoting the deployment of WiMAX and its related technologies as an alternative source of residential broadband to cable TV and DSL.

---

[9]Michael K. Powell, "WISPs: Bringing the Benefits of Broadband to Rural America" (remarks as prepared for delivery at WISPCON, Las Vegas, NV, October 27, 2004).

11

How to Dismantle a PSTN: The Business Case for WiMAX

Overview

The preceding ten chapters have discussed the technology of WiMAX and a number of potential applications. The purpose of this chapter is to determine the big "so what?!" of WiMAX. This chapter will examine the WiMAX applications that exist and how they can potentially disrupt existing service and power structures in the telecommunications industry.

Immediate Markets

Where can WiMAX be applied today to save money on both operating expenses (OPEX) and capital expenditures (CAPEX)? The following sections describe where and how WiMAX can save money and open new markets.

Local Loop Bypass Many businesses in the small-to-medium enterprise (SME) market pay dearly every month for what is billed as "local loop," the charge for transporting data over copper wire to the customer's premise. This charge applies to the firm's data T1 (1.54 Mbps) or voice T1 for circuit-switched voice. This cost is often hundreds of dollars per month per T1. For DS3s (45 Mbps) the cost per month may be in the thousands of dollars. The farther the customer's premises are from the telephone company's central office, the greater the cost. By adopting WiMAX as a local loop bypass, businesses can save money two ways:

1. They can eliminate or greatly reduce the monthly local loop fee by getting their T1 or DS3 data from a WiMAX-enabled service provider.

2. They can eliminate or greatly reduce the monthly cost on the local loop charge for their voice T1 service by switching to a VoIP service provider and utilizing WiMAX as an alternative to the local loop copper.

Figure 11-1 illustrates bypassing local loop charges by using WiMAX as a local loop alternative. Figure 11-2 details savings on local loop charges when utilizing WiMAX as a VoIP delivery alternative to the telephone company's circuit-switched telephone service.

Residential and SOHO High-Speed Internet Access Today this market segment is primarily dependent on the availability of DSL or cable. In some areas the available services may not meet customer expectations for performance or reliability and/or are too expensive. In many rural areas, residential customers are limited to low-speed dial-up services. In developing countries, many regions have no available means for Internet access. The analysis will show that the WiMAX technology will enable an operator to economically address this market segment and have a winning business case under a variety of demographic conditions.

Small and Medium Business This market segment is very often underserved in areas other than highly competitive urban environments. The WiMAX technology can cost-effectively meet the requirements of small and medium size businesses in low-density environments and can also provide a cost-effective alternative in urban areas competing with DSL and leased line services.

Figure 11-1
WiMAX is an alternative to the telephone company's local loop charges on data circuits and can save the enterprise on monthly local loop charges.

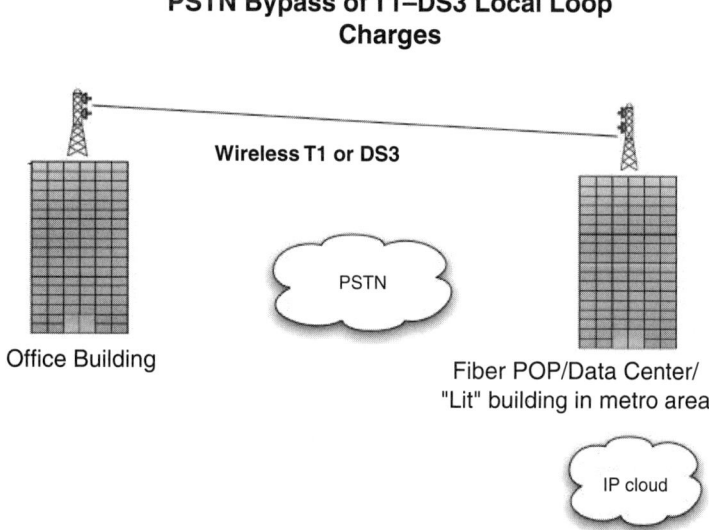

PSTN Bypass of T1–DS3 Local Loop Charges

Wireless T1 or DS3

PSTN

Office Building

Fiber POP/Data Center/ "Lit" building in metro area

IP cloud

Figure 11-2
Savings on local
loop charges
when WiMAX is
used as a VoIP
delivery
alternative

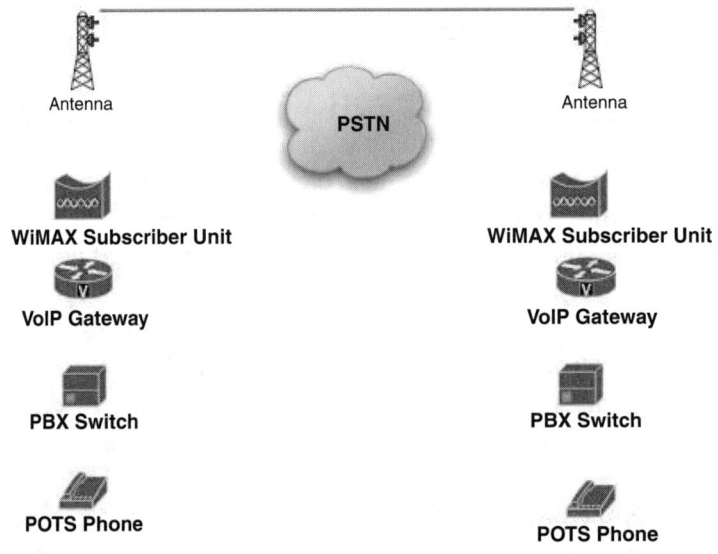

Wi-Fi Hot Spot Backhaul   Wi-Fi hot spots are being installed worldwide at a rapid pace. One of the obstacles for continued hot spot growth, however, is the availability of high capacity, cost-effective backhaul solutions. This application can also be addressed with the WiMAX technology. And with nomadic capability, WiMAX can also fill in the coverage gaps between Wi-Fi hot spot coverage areas. Figure 11-3 illustrates WiMAX as a backhaul to existing Wi-Fi networks.

Secondary Markets

The following applications are not included in the business case analysis. Nevertheless, they are worthy of mention, as they represent additional potential revenue sources for the wireless operator.

Cellular Backhaul   In the United States, the majority of backhaul is done by leasing T1 services from incumbent wire-line operators. With the WiMAX technology, cellular operators will have the opportunity to decrease their independence on backhaul facilities leased from their competitors. Outside the United States, the use of point-

Figure 11-3
WiMAX as backhaul supporting Wi-Fi networks

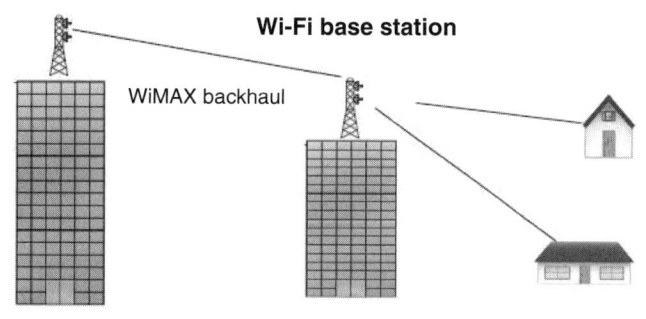

WiMAX base station

Wi-Fi base station

WiMAX backhaul

Wi-Fi subscribers

to-point microwave is more prevalent for mobile backhaul, but WiMAX can still play a role in enabling mobile operators to cost-effectively increase backhaul capacity using WiMAX as an overlay network. This overlay approach will enable mobile operators to add the capacity required to support the wide range of new mobile services they plan to offer without the risk of disrupting existing services. In many cases, this application will be best addressed through the use of WiMAX-based point-to-point links sharing the PMP infrastructure.

Public Safety Services and Private Networks Support for nomadic services and the ability to provide ubiquitous coverage in a metropolitan area provide a tool for law enforcement, fire protection, and other public safety organizations, enabling them to maintain critical communications under a variety of adverse conditions. Private networks for industrial complexes, universities, and other campus type environments also represent a potential business opportunity for WiMAX.

Demographics

Demographics play a key role in determining the business viability of any telecommunications network. Traditionally, demographic regions are divided into urban, suburban, and rural areas. In our analysis, a fourth area, called exurban, has been added. Exurban areas are primarily residential and compared to suburban areas are further from the urban center with lower household densities. DSL availability is limited because of the distance between the end-user and the switching center, and cable, in many cases, is simply too expensive.

Rural areas for the purpose of the business case analysis are defined as small cities or towns that are located far from a metropolitan area. Customer densities can be fairly high in these areas, but they tend to be underserved because of their remote location. Table 11-1 summarizes the characteristics that will generally be encountered in each of the four geographical areas under consideration for a new wireless service provider.

Services

A description of the services used in the business case with the assumed first year annual revenues per user (ARPUs) follows. These ARPUs are competitive with or below current cable, DSL, and leased

Table 11-1

Demographic Market for WiMAX

Area	Characteristics
Urban	▪ highest density potential WiMAX subscribers
	▪ many multiple tenant office and residential buildings
	▪ smaller WiMAX cell sizes to meet capacity requirements
	▪ strong competition driven by market size and availability of alternate access technologies
	This competitive environment leads to
	▪ lower market penetration
	▪ higher marketing and sales expense
	Other considerations include
	▪ licensed spectrum a good idea: minimize potential for interference

Area	Characteristics
Suburban	▪ moderate density of potential WiMAX subscribers
	▪ higher percentage single family residences
	▪ business parks, strip malls
	▪ cable or DSL may not be widely available
	▪ higher market penetration for new operator
Exurban	▪ upscale residential neighborhoods with moderate to low household density
	▪ fewer businesses
	▪ high concentration of computer users
	▪ cable/DSL not widely available
	▪ larger WiMAX cell sizes, possible terrain and range limitations
	▪ BS development costs impacted by environmental impact studies, architectural reviews, and so on
	▪ high percentage of commuter need for telecommuter services
	▪ high market penetration expected for fixed BS Internet access
Rural (small, relatively isolated cities and towns)	▪ market is residential and small business
	▪ little if any cable/DSL
	▪ high pent-up demand for Internet access
	▪ limited competition
	▪ very high market penetration and rapid adoption rate expected for new operator
	▪ high capacity (DS3) backhaul may be a challenge

line services in most developed countries. For the business case analysis, the ARPUs are assumed to drop 5 percent per year after the first year. Wire-line operators generally offer several types of services for SME, but for the sake of simplicity, only two service levels have been assumed for this analysis.

In addition to high-speed Internet access, it is assumed the operator will also offer voice services to residential and SME customers. Other revenue sources include one-time activation fees and equipment rental fees for operator-supplied customer premise equipment. These fees are assumed to stay constant over the business case period. Regulator imposed taxes and tariffs are not included in the analysis because these costs are generally passed through to the end-customer and will, therefore, have little or no impact on the business case.

Frequency Band Alternatives

A key decision regarding spectrum choice is whether to use licensed or unlicensed spectrum. The use of licensed spectrum has the obvious advantage of providing protection against interference from other wireless operators. The disadvantage is dealing with the licensing process. This process varies depending on local regulation. It can be very simple and quick or complex and lengthy, and in countries where auctions are used, it can be expensive in highly sought-after regions. The use of unlicensed spectrum gives the wireless operator the advantage of being able to deploy immediately but runs the risk of interference from neighboring wireless operators in the future. In general, our feeling is that the use of licensed spectrum is desirable in major metropolitan areas where multiple wireless operators are more likely.

License-exempt spectrum, on the other hand, is often a good choice in rural areas where fewer operators are likely to exist. In these areas, interference mitigation is easily accomplished through frequency coordination between the operators. A good practice when deploying with unlicensed spectrum is to size hubs so that no more than half the available band is used. This enables the use of automatic channel selection to enable auto-selection of channels that are not subject to interference from other wireless operators.

The frequency bands that are of primary interest with today's prevailing regulations are:

■ The license-exempt 5.8 GHz, known as Universal National Information Infrastructure (UNII) Band in the United States

■ The licensed 2.5 GHz, known as Multipoint Distribution Service (MDS) Band, aka Broadband Radio Service (BRS) in the United States

■ The licensed 3.5 GHz band or the licensed-at-no-cost 3.65 GHz band (United States only)

A summary of these bands and relevant considerations for the WiMAX business case is provided in Table 11-2. In our analysis, we will use the 3.5/3.65 GHz band for metropolitan area deployment and the 5.8 GHz unlicensed band for rural area deployment.

Table 11-2

The Business Case for the WiMAX Operator

Customer	Service	Other Revenue	Monthly Revenue
Residential Data + VoIP	A "best effort" service (assume 384Kbps with 20:1 over-subscription) $30 + $25/month VoIP	$10/month for equipment lease/ one-time $50 service activation fee	$65 ($30 service + $25 VoIP + $10 lease)
Small to Medium Business (SMB)	1.0 Mbps CIR, 5 Mbps PIR @ $450 VoIP @ $50/line/month (for example, 10 lines = $500)	$35/month equipment lease fee and one-time $500 service activation	$985 ($450 + $500 + $35, see Service column for details)
Wi-Fi Hot Spot Backhaul	1.5 Mbps CIR, 10+ Mbps PIR	$25/month equipment lease + $500 activation fee	$675 ($650/month service + $25/month equipment lease)

Geographic Scenarios for Business Case Analysis For the business case analysis, three different scenarios are analyzed; the characteristics of these scenarios are summarized in Table 11-3.

Capital Expense (CAPEX) Items

What makes WiMAX a disruptive technology? One explanation is a low barrier to entry due to the relative (to copper or fiber optic) low cost of infrastructure.

Table 11-3

Summary of Business Case Scenarios

Element	Scenario 1	Scenario 2	Scenario 3
Geographic area	City/ metro area	City/ metro area	Rural/small town
Market segment	Residential	Residential /SME/Wi-Fi backhaul	Residential/SME
Size	125 sq km	125 sq km	16 sq km
Population	1 million	1 million	25,000
Residential density	6,000 homes/sq km (urban); 1,500 homes/sq km (suburban); 500 homes/sq km (exurban)	6,000 homes/sq km (urban); 1,500 homes/sq km (suburban); 500 homes/sq km (exurban)	600 households/sq km
Total homes	390,000	390,000	9,600
Total SME	N/A	24,000	N/A
Adoption rate	4 years	4 years	3 years
Frequency band	3.5 GHz (licensed)	3.5 GHz (licensed)	5.8 GHz (unlicensed)
Channel BW	3.5 MHz FDD	3.5 MHz FDD	10 MHz TDD
Assumed spectrum	28 MHz (2×14 MHz)	28 MHz (2×14 MHz)	60 MHz

BS Edge and Core Network   The business case assumes a green field deployment, and as such, it must include an allowance for core and edge network equipment in addition to WiMAX-specific equipment (see Figure 11-4). Most of this equipment must be in place prior to offering services. BSs and BS equipment need not be installed in totality at the outset but can be deployed over a period of time to address specific market segments or geographical areas of interest to the operator. Nevertheless, in a metro area, it is desirable to install a sufficient number of BSs to cover an addressable market large enough to quickly recover the fixed infrastructure costs. It is also desirable in the case of fixed services involving operator-installed outdoor CPEs with directional antennas to locate and deploy BSs in such a way so as to minimize the possibility of having to insert other BSs within the same coverage area to add capacity. This approach would generally require potentially expensive truck-rolls to redirect outdoor CPE antennas and can be avoided with careful long-range market analysis and RF planning. If sufficient spectrum is available, BS capacity can be increased by simply adding additional channels to all or to selected BSs as required to match BS capacity to growing customer requirements. This is an ideal way to phase the deployment and grow the wireless network capacity to match customer growth. In the business case analysis, BS capacity is determined by using a 20:1 over-booking factor for best-effort residential services assuming 384 Kbps average data rate and 1:1 for SME committed information rate (CIR) services. For the residential case this conservative over-booking factor should enable WiMAX subscribers to experience performance during peak periods superior to what many DSL and cable customers experience today. In scenarios 1 and 2, it is assumed that all the BSs necessary to meet long-term capacity requirements would be deployed prior to offering services. In scenario 3, a single BS is deployed to cover the region, and two channels are added in year 3 to increase capacity. In very large metropolitan areas an operator may choose to deploy BSs over several years to spread out the capital investment by dividing the area into smaller geographic regions and fully covering one region prior to moving on to the next.

The business case also assumes the deployment of a high capacity point-to-point wireless backhaul connection for each BS to a point of

Figure 11-4
Bypassing cable
TV infrastructure
with WiMAX

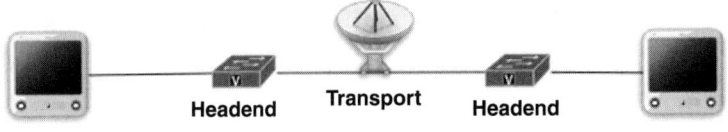

Legacy Cable TV

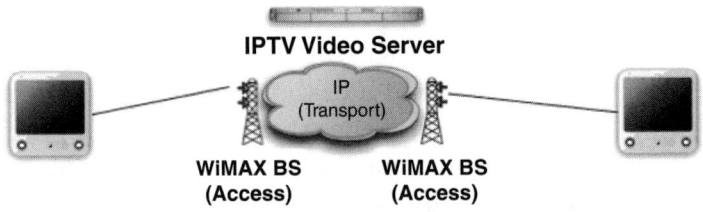

Cable TV Bypass with WiMAX and IPTV

presence or fiber node for connection to the core network. This can also be accomplished by means of leased T1/E1 lines. In this case, there would be an operating rather than capital expense. Table 11-4 summarizes the BS and infrastructure costs that have been assumed for the three business case scenarios. For scenarios 1 and 2, it is assumed that a spectrum license is obtained through an auction process at a cost of $.01 per MHz pop5. In some countries, licenses can be obtained at no initial cost but with an annual lease fee. In these cases, the cost to the operator would be entered as an operating rather than capital expense. Table 11-4 provides an overview of capital expenditures necessary to deploy WiMAX.

CPE Equipment

WiMAX equipment manufacturers will be providing CPE hardware in a variety of port configurations and features to address the needs of different market segments. Residential CPEs are expected to be available in a fully integrated indoor self-installable unit as well as an indoor/outdoor configuration with a high-gain antenna for use on customer sites with lower signal strength. In the business case analysis, a percentage breakdown of each is assumed in accordance with the frequency band, cell radius, and propagation conditions

Table 11-4

CAPEX for Network Infrastructure

Description	Scenario 1	Scenario 2	Scenario 3	Comments
WiMAX equipment	$8K/BS (3 sector config)	$8K/BS (3 sector config)	$8K/BS (3 sector config)	Add $1K/ additional sector
Other BS equipment	$10K	$10K	$10K	Cabinets, network interface cards, and so on
Backhaul link	$10K Pt-to-Pt microwave link	$10K Pt-to-Pt microwave link	$100K	One multiple hop for rural area
Core and edge equipment	$200K	$250K	$50K	Router/ATM switch/NMS
Spectrum license	Assume $.01/MHz /POP	Assume $.01/MHz /POP	N/A	License acquired as upfront investment
BS acquisition and civil works	$50K average	$50K average	$50K average	Indoor/outdoor site preparation, cabling, and so on

that are likely to be encountered in the different geographical areas. CPEs for SME will generally be configured with T1/E1 ports in addition to 100BT Ethernet ports. These units are priced higher for the business case, consistent with the added performance.

For both the residential and SME market segment, it is assumed that a percentage of customers will opt to supply their own equipment rather than pay an equipment lease fee to the operator. This has the effect of reducing the CPE CAPEX and CPE maintenance expense. It also, however, reduces operator revenues derived from equipment lease fees. Because of this interrelationship, the impact on the payback period is not significant.

The business case analysis assumes that the price of residential terminals will drop by about 15 percent per year due to growing volumes and manufacturing efficiencies, and lower volume business terminals will drop by about 5 percent per year. The CPE costs used in the business case analysis are summarized in Table 11-5.

Table 11-5

Assumptions
Regarding CPE

CPE Type	Year 1 CAPEX	Annual Price Reduction	Assumes Percent of CPEs Provided by Operator
Residential Indoor Self-Installed CPE	$250	15 percent	80-percent scenario 1, 60-percent scenario 2
Residential Outdoor CPE	$350	15 percent	See above
Small Business Terminal	$700	5 percent	50 percent
Medium Business Terminal	$1,400	5 percent	50 percent
Wi-Fi Hotspot Terminal	$300	5 percent	20 percent

Operating Expense (OPEX) Items

The OPEX items used in the business case analysis are summarized in Table 11-6.

The Business Case

What markets can WiMAX be applied to? What is the business case for WiMAX in that market? The following sections will explore applications and markets for WiMAX.

Scenario 1: Residential Market Segment in a Metro Area Environment

A market financial summary for this scenario has been provided in Table 11-7. The spectrum available to the operator is assumed to be limited to 28 MHz (2 × 14 MHz). The WiMAX BS equipment uses 3.5 MHz channels with frequency division duplexing. A four-sector BS, therefore, can be deployed using one channel pair per sector. Due to the limited spectrum, the BSs in each of the three geographical areas are capacity-limited rather than range-limited, and 26 BSs are required to provide services to 6.3 percent of the addressable resi-

Table 11-6

OPEX
Considerations

OPEX Item	Business Case Assumption	Comments
Sales/marketing expense (includes customer technical support)	20 percent of gross revenue in year 1, 11 percent year 5	Higher percent of revenue in early years to reflect fixed costs associated with these expenses, fifth year levels consistent with levels of a mature stable business
Network operations	10 percent of gross revenue in year 1, dropping to 7 percent in year 5	See above
G&A	6 percent of gross revenue in year 1, dropping to 3 percent in year 5	See above
Equipment maintenance	5 percent of CAPEX for BS gear, 7 percent of operator-owned CPE	Reflects higher maintenance costs associated with maintaining remotely located equipment
BS installation and commissioning	$3K for a 4 sector BS	One-time expense
CPE install	Varies	Offset: install charge to subscriber
BS site lease	$1,500/month/BS	Space for indoor equipment plus antenna space lease
Customer site lease	$50/month	Does not apply to residential market
Bad debt and churn	12 percent residential and 3 percent SME	

dential market. With a 4-year market adoption rate to reach 90 percent of the target market penetration, installation and commissioning costs peak in years 3 and 4. This contribution to OPEX plays a lesser role in the 5th year, as the annual rate of customer growth slows.

Table 11-7

Market
Summary for
Scenario 1

Spectrum		Deployment Data	
Frequency band	3.5 GHz	WiMAX BS deployed	26
Channel BW in MHz	3.50 GHz	Aggregate payload in Mbps/sq km	1,005
Spectrum required in MHz	28	Coverage area sq km	125
Addressable market		Average data density Mbps/sq km	8
Households covered	388, 254	Population in coverage area	1,009,481
Businesses covered	N/A	**Assumed CPE Mix**	
Market penetration (5th year)		Percent of indoor residential CPE	80 percent
Market adoption curve	4 years	Percent of residential CPEs operator-supplied	80 percent
Residential market	6.3 percent	Percent SME CPEs operator-supplied	N/A
Residential voice services	23 percent	ARPU price erosion	5 percent
SME market	N/A	Average number subs/BS	948
SME voice	N/A	CAPEX/subscriber	5,328
Wi-Fi hotspot backhaul	N/A	Total CAPEX in $M	$8.1
Net present value (5 years) millions	**$3.6**	**IRR**	**90 percent**

The CAPEX is dominated by WiMAX CPEs because it is assumed that the operator would provide 80 percent of the equipment for this scenario. This, of course, is offset by the $10 per month equipment rental fees. As CPE prices decline, we would expect a higher percentage of CPEs to be purchased by the customer to avoid the rental expense. With an internal rate of return (IRR) of 90 percent, this is clearly an attractive business model.

Future Markets

Applications for WiMAX are limited only by the imagination of the entrepreneur. The following sections explore some relatively simple applications.

Replacing Cell Phone Infrastructure   Figure 11-5 details how a cell phone network can be bypassed utilizing WiMAX infrastructure. Table 11-8 lists cost savings of WiMAX versus cell phone infrastructure. According to Christensen's *Innovator's Dilemma*, disruptive technology is defined as being "cheaper, simpler, smaller and more convenient to use." Note the cost differences among the platforms contained in the infrastructure of the two network types.

Bypass by Substituting for the PSTN   Assuming the process shown in Figure 11-5 became the standard practice for bypassing the cell phone network, what would be the demand for a "land line" telephone? The convenience of a mobile telephone offered at a cost competitive to that of the legacy land line could drive the copper wire-connected, circuit-switched telephone into extinction. Table 11-9 compares infrastructure costs. Given the lower barrier to entry presented by WiMAX, it is not hard to imagine a number of entrepreneurial companies seeking to take away market share from incumbent telephone companies.

Figure 11-5
WiMAX as cell phone bypass

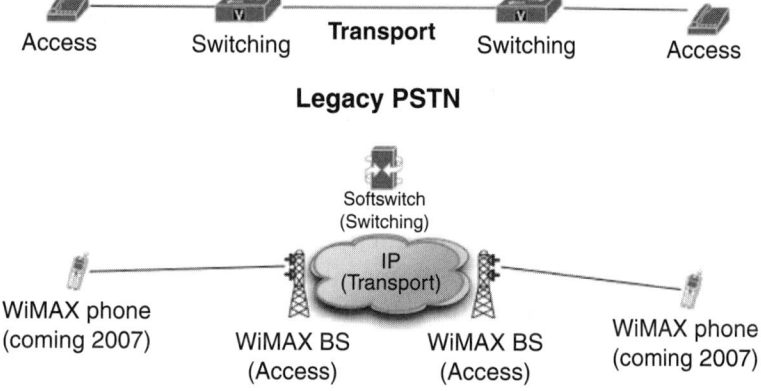

Access   Switching   **Transport**   Switching   Access

Legacy PSTN

Softswitch
(Switching)

IP
(Transport)

WiMAX phone
(coming 2007)

WiMAX BS
(Access)

WiMAX BS
(Access)

WiMAX phone
(coming 2007)

PSTN Bypass with WiMAX and VoIP

Table 11-8

Replacing Cell
Phone
Network with
WiMAX

Cost Component	Legacy Cell	WiMAX
Switching	Class 4 and 5 switches at $10 million each (need several to cover diverse geographic footprint)	Softswitch at $500,000 each (need one, buying licenses and servers to scale and for redundancy)
Access	Expensive BSs; very expensive spectrum in most markets	Inexpensive BSs; unlicensed spectrum is free (Ex. 20 channels 5.2, 5.4, and 5.8 GHz)
Transport (backhaul)	Uses expensive RBOC DS3 and T1	WiMAX as backhaul; with unlicensed spectrum, only cost is radios at $1,500 each
Revenue stream	Mostly voice, limited data	High bandwidth allows voice (fixed and mobile), video, and data

Replacing or Competing with Cable TV Infrastructure If the PSTN's copper wire infrastructure could be bypassed by WiMAX, can the cable TV company's coaxial cable infrastructure be at equal risk from bypass by WiMAX? As Figure 11-4 and Table 11-10 outline, it is certainly possible.

Can it make money for the service provider? Does it present a significant lowering of barriers to entry to the broadband Internet market? The absence of cabling and obtaining rights-of-way would be the first indication of potential savings in the installation of a network. Perhaps one of the strongest arguments in favor of WiMAX is that it potentially presents a cost-effective means of offering broadband Internet service to a mass market with the least expense in infrastructure relative to wired technologies (twisted pair copper, coax cable, fiber-to-the-home). This low cost in infrastructure promotes the deployment of WiMAX services by less well-capitalized entrepreneurs, municipal networks, and even "free net" community networks built and maintained by volunteers. The growth of WiMAX networks is often described as being "viral," that is, unplanned or "grassroots."

Table 11-9

Cost
Comparison:
PSTN vs.
WiMAX

Cost Component	Legacy PSTN	WiMAX
Switching	Class 4 and 5 switches at $10 million each (need several to cover diverse geographic footprint)	Softswitch at $500,000 each (need one, buying licenses and servers to scale and for redundancy)
Access	Uses copper wire requiring expensive right-of-way for wiring, poles, repeaters, pedestals, and so on	Inexpensive BSs; unlicensed spectrum is free; except for roof and tower rights, little need for right-of-way
Transport (backhaul)	Uses expensive RBOC DS3 and T1; fiber optic cable requires trenching and right-of-way	WiMAX as backhaul; with unlicensed spectrum, only cost is radios at $1,500 each
Revenue stream	Voice, low bandwidth data	High bandwidth allows voice (fixed and mobile), video, and data

Table 11-10

Cost
Comparison:
Cable TV vs.
WiMAX

Cost Component	Legacy Cable TV	WiMAX
Switching	Video: Expensive headends Voice: Class 4 and 5 switches at $10 million each (need several to cover diverse geographic footprint); some players using VoIP (softswitch)	Video: Cable TV programming available via Voice IPTV: Softswitch at $500,000 each (need one, buying licenses and servers to scale and for redundancy)
Access	Expensive coax and cable; expensive right-of-way	Inexpensive BSs; unlicensed spectrum is free
Transport (backhaul)	Uses expensive satellite (hundreds of millions of dollars to build and launch) or fiber	WiMAX as backhaul; with unlicensed spectrum, only cost is radios at $1,500 each

Economics of Wireless in the Enterprise

The economics of WiMAX in enterprise applications should be assessed in two ways: first, comparing applications where the wireless network is simply less expensive to deploy than the wired network where both applications perform the same function and, second, analyzing situations where a wireless network enables employees to be more efficient. Money saved is money earned.

You *Can* "Take It with You When You Go"

WiMAX as a technology could gain wide acceptance in enterprise networks. The reasons for this are many including cost savings, mobility, and employee productivity. The origin of wireless networks lies in the convenience of not having to run Category 5 or telephone wiring in an enterprise environment. The cost of the wire itself is not so great; however, the labor to perform the installation and the boring of holes in walls and other defacing of property necessary to run the wire runs up the cost of a wired LAN as compared to subscribing to a wireless broadband service.

A timeless wisdom regarding death and personal wealth goes "You can't take it with you when you go." Most commercial lease agreements in North America hold a proviso that wired infrastructure must remain in the building when the enterprise tenant vacates the premises (most do so for more advantageous rent). This is a sunk cost that the enterprise tenants lose when they move to another building space. In contrast, the WiMAX broadband Internet service is almost completely portable. The deployment of a wireless enterprise network allows the enterprise greater flexibility when shopping for more advantageous rents. Table 11-11 offers a brief outline of savings for wireless versus wired office space.

The WiMAX/Wi-Fi Wireless Office Significant savings can be achieved by moving the office from "wired" to "wireless." Figure 11-6 illustrates how an office could receive its data from a WiMAX

Table 11-11

Cost
Comparison:
Installation of
Wired LAN vs.
Wireless LAN

Cost Component	Cost per Unit	Required Wired Network Units	Total Cost for a LAN	Required Wired Network Units	Total Cost for a WLAN
Cisco 1721 router	2,000	1	2,000	outsourced	
Cisco 3524 switch	2,000	1	2,000	outsourced	
Dell server	2,500	1	2,500	outsourced	
Laptop w/ built-in Wi-Fi	1,500	10	15,000	10	15,000
Desktop Wi-Fi card for PC	1,000	1	1,000	1	1,000
Printer	2,000	1	1,000	1	1,000
Wi-Fi access pts	500	0	0	1	500
VPN/ encryption	1,500	0	0		0
T1	500 wired/ 200 from WiMAX provider	1	500	1	200
Installation CAT5 wire drops/runs	250	10	4,000		0
Telephone key system with handsets	5,000	1	5,000		
Wi-Fi telephone handsets	150			10	1,500
Totals			35,000		19,200

Note: WLAN equipment pricing may fall faster than LAN gear as technology matures.

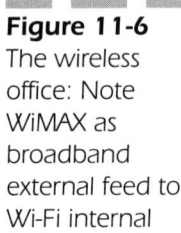

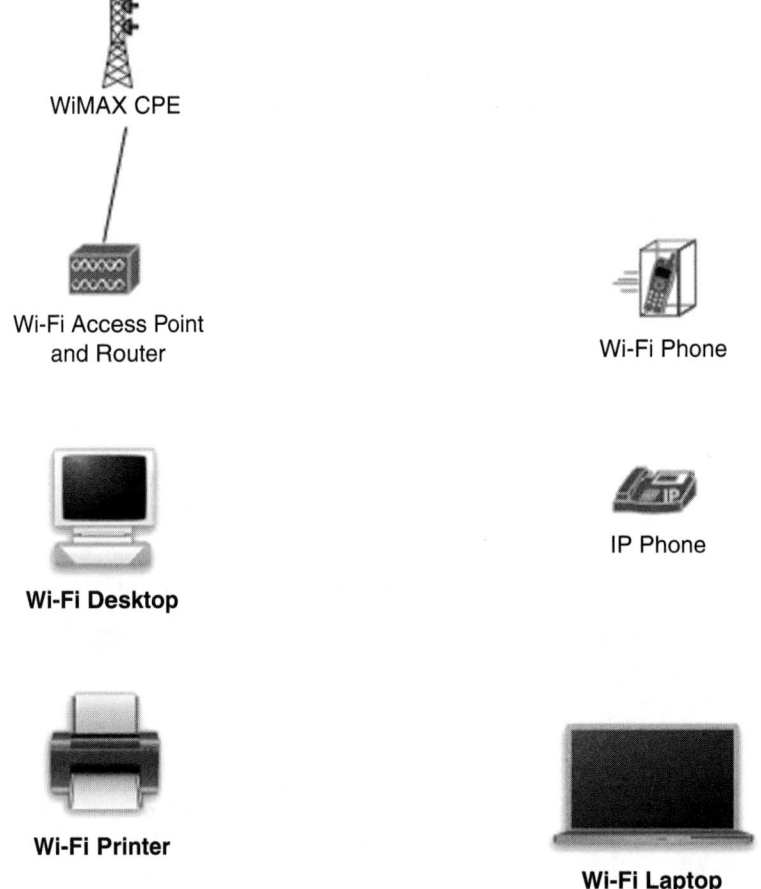

WiMAX CPE

Wi-Fi Access Point
and Router

Wi-Fi Phone

Wi-Fi Desktop

IP Phone

Wi-Fi Printer

Wi-Fi Laptop

source. Office components such as computers, telephones, and print-
ers can all be networked via wireless means. Refer to Table 11-11 for
a comparison of cost components of both the wired and wireless
office.

Economics of WiMAX in Public Networks

What is the economic pull to grow wireless public networks? The previous section of this chapter described the advantages of WiMAX in
private networks. How then will public networks become accepted in

our economy? In an ideal world, some form of ubiquitous wireless coverage would extend to at least every residence and small business in a metropolitan area. From that goal, extending the coverage to small towns and farms could occur at a rapid pace, assuming a business model propels that growth.

Although the Telecommunications Act of 1996 was intended to bring competition to the local loop, some six years after its passage, fewer than 10 percent of United States residences enjoy any choice in their local telephone service provider. Competition will never come *in* the local loop but rather *to* the local loop. The act prescribed a formula for competitors to lease facilities (copper wire and switch space) from incumbent service providers. One of the reasons competition in the local loop is lacking is simply the cost of deploying competing strands of copper wire.

According to FCC studies, the cost to install copper loop plant depends on the density of households in the service area. This cost can range from $500 per household in the least expensive urban sites to a typical $1,000 in dense suburban areas, ascending to $10,000/loop in outlying rural areas. Economies of scale apply here. A competitor cannot come close to matching incumbent costs on loop plant because a competitor with a low market share has, effectively, rural density (and costs), even in an urban area.[1] Competitors to an incumbent telephone company must, then, consider their return on investment (ROI) on a per customer basis. If the competitor will realize $40 per month on a customer, for example, the ROI period could be very long. If a wireless service provider could persuade the customer to purchase his or her own customer premises equipment (CPE), the wireless competitor could potentially be more competitive than any other form of competitive service compared to the incumbent telephone company.

Advantages to SOHO or Residence   Earlier scenarios detailing cost savings for WiMAX service providers carry through to residential subscribers as well. Table 11-12 outlines savings for residential subscribers who have their services (converged) from one wireless

---

[1] Fred Goldstein (telecommunications consultant), interview, November 28, 2002.

Component	Conventional	WISP
Local phone service (per line)	$25	$20 (VoIP service provider)
Long distance	$100 ($.07/minute)	$0 (assuming all calls VoIP)
Video (cable vs. video on demand)	$50	$0
Internet	$25	$0
Broadband device (DSL, cable)	$40	$45
TOTALS	$240	$65

ISP (WISP) as opposed to buying those services separately from diverse service providers.

Economic Benefits of Ubiquitous Broadband

A wave of opportunity for wireless broadband applications is in the making. Most of it lies in the form of broadband deployment. In their April 2001 white paper, "The $500 Billion Opportunity: The Potential Economic Benefit of Widespread Diffusion of Broadband Internet Access," Robert Crandall and Charles Jackson point to an economic benefit of $500 billion per year for the American economy if broadband Internet access were to be as ubiquitous as land line phones. Given that WiMAX makes deployment of residential broadband much less expensive, the following pages will outline the benefits of ubiquitous WiMAX deployment.

In their 2001 report, economists Crandall and Jackson explored the benefits to the United States economy if broadband Internet were to become as widespread as telephone service is today. The remainder of this chapter assumes that it is considerably less expensive (both in terms of hardware and lawyers) to deploy wireless broadband Internet to a residence than a similar service that depends on wiring (copper wire from the phone company or coax cable from the cable TV company). Both telephone wires and cable TV coax cable run by (are accessible by) almost 90 percent of American homes. The physical cost of connecting a home to the Internet in most residential applications is not that high. However, for a new market entrant, gaining the right-of-way from private land owners and public utilities to get to those households will not be possible in most cases without costly legal procedures. Revenue generated from subscription fees may not offset the legal costs of running wire or cable to that residence.

Using WiMAX as a means of access does not require legal dealings for rights-of-way and, relative to wired infrastructure, can be deployed much more quickly. As evidenced by the efforts of CLECs to offer competitive residential telephone service using incumbent telephone poles and other incumbent-owned and incumbent-operated facilities, it is far easier to bypass PSTN facilities than to utilize them via legal means. A wireless service provider need only install a BS and turn up service. The remainder of this chapter will explore the benefits of ubiquitous residential broadband Internet access, assuming the ease and economy of WiMAX is a catalyst for achieving the same levels of penetration for broadband Internet access as residential telephone service has today.

As the uses of broadband multiply, the value to subscribers rises far above the monthly subscription price. This is the *consumer surplus* from the innovation. Producers of new services that rely on broadband (see example of i-mode-type services, Net2Phone, and so on), of products used in conjunction with broadband service (softswitches, media gateways, IP phones, residential gateways), and even of the broadband service itself also gain from the greater diffusion of broadband. The *producer surplus* that is generated by sales is a real benefit to producers and, therefore, to the economy. At present, no more than 8 percent of American households subscribe to a broadband service; only slightly more than 50 percent subscribe to an Internet service of any kind; and 94 percent subscribe to ordinary telephone service.[2] Were broadband to become ubiquitous, it would resemble current telephone service in its household penetration.

Producer Benefits   Figure 11-7 demonstrates the economic pull-through of wireless broadband.

One of the reasons many IP backbone and wireless local loop carriers went bankrupt is that they could not deliver bandwidth to a broad market. The "bottleneck" to the last mile remains the access controlled largely by telephone companies with their ubiquitous twisted-pair copper wire. Cable TV companies now service a majority of American homes. WiMAX presents a means of reaching customers anywhere and everywhere with minimum cost to the service provider.

WiMAX will create a cycle of adoption that will drive technology purchases and upgrades by enterprises, retailers, service providers,

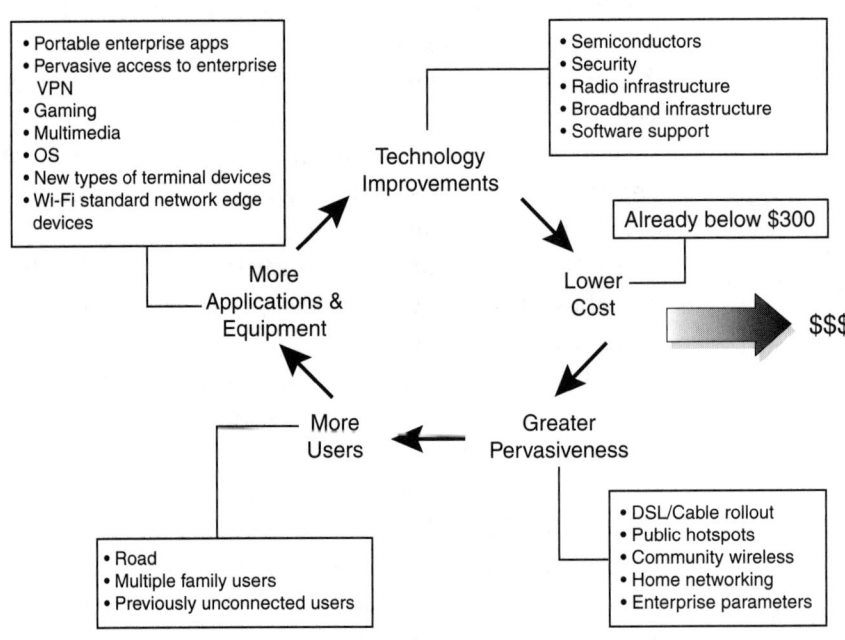

Figure 11-7

Economic pull-through of wireless broadband (Figure courtesy Goldman Sachs)

---

[2]The number of broadband subscribers (DSL plus cable modems) was 7.3 million as of March 2001. *See* "Failure of Free ISPs Triggers First-Ever Dip, to 68.4 Million Online Users: Cable Modem Boom Continues, as DSL Sign-ups Lag," *Telecommunications Reports,* April 2001. The estimates for Internet and telephone service are from authors' tabulations using the Current Population Survey for August 2000.

government, and individuals for the following three reasons. First, it offers a means of delivery that is "cheaper, simpler, smaller, and more convenient" than wired (telephone and cable TV) means of delivery. WiMAX service requires either the presence or installation of a BS and customer premise equipment. The larger the environment (the number of coverage areas, number of users supported, and so on), the more infrastructure equipment and network bandwidth are required, thus spurring sales of BSs, customer premise equipment, and so on.

Once WiMAX is available in a given area, it will spur the purchase of more mobile computers, PDAs, pocket PCs, and other wireless devices. This is particularly relevant in the home, where WiMAX enables broadband connections to be shared easily among multiple PCs and, ultimately, other devices as well. Major PC vendors will soon include WiMAX support. Chipmakers and laptop computer manufacturers will soon offer WiMAX capabilities in their products. These market drivers include home networking, home multimedia, smart appliances, and VoIP. These applications require new platforms such as home access points and voiceover WiMAX telephony devices.[3]

Computer Sales   The expansion of the demand for broadband will create additional demand for computers and networked home appliances. Approximately 40 percent of all United States households do not currently have a computer.[4] These households are clearly not equipped to connect to the Internet at any speed. Of the 60 percent of households with computers, many will need to upgrade their equipment to obtain greater processing speed, more random-access memory, or greater hard-drive capacity. Still others will choose to buy more advanced equipment such as storage devices, MP3 players for music downloads, and LCD projectors for viewing video downloaded via a high-speed Wi-Fi or WiMAX connection. Applications

---

[3] Chris Fine, "Watch Out for Wi-Fi," white paper from Goldman Sachs.

[4] The most recent estimate from the Bureau of the Census for June 2000 was 41.5 percent. More recent estimates from TNS suggest that about 50 percent of households now have access to the Internet. *See* TNS Telecoms, *ReQuest Market Monitor National Consumer Survey,* vol. 3 (2001).

(video, telephony) that the following pages will explore could very well drive much of the remaining 40 percent of households without computers to make the leap and install a computer in their homes.

Crandall and Jackson estimate broadband's stimulus on household purchases of broadband-related equipment would be that United States household spending on computer equipment, peripherals, and software would resume its 1991–1995 rate of growth of 14.3 percent per year, rather than continuing at its 1995–1999 growth rate of 10.4 percent per year. If growth returns to its 1991–1995 pace, by 2006 total spending would be $80 billion, rather than $66 billion—an increase of $14 billion. By 2011, the difference would be $53 billion per year. Were the broadband revolution to accelerate household equipment expenditures by another 3 percent per year to 17.3 percent annual growth, the additional spending in ten years would be $110 billion per year.[5]

Consumer Benefits   The most straightforward estimate of the value of enhanced availability of broadband derives from information on consumer subscriptions to broadband services.

An Estimate Based on Current Demand   Price elasticity of demand is a relationship of change in demand to the change in price. Given current broadband penetration of 8 percent and an average price of the service of $40 per month, total broadband revenues may be estimated at $480 times 8.4 million or $4 billion per year. Assuming that the demand for such service is linear with an elasticity of −1.0, the value of the service to these consumers—the consumer surplus—is $2 billion per year in addition to the $4 billion they pay. If the demand elasticity is −1.5, the consumer surplus falls to $1.4 billion.

Were broadband to spread to 50 percent of households at $40 per month through a shift of a linear demand curve with constant slope, the annual expenditure on the service would rise to $31.2 billion. At 50 percent penetration, the additional value to consumers would rise

---

[5]Robert Crandall and Charles Jackson, "The $500 Billion," white paper from Criterion Economics LLC.

to between $80 billion and $121 billion per year at these two price elasticities. If broadband service were to become truly ubiquitous, similar to ordinary telephone service, annual consumer expenditures on the service would rise $58.7 billion per year, assuming the continued shift of the linear demand curve at constant slope and an annual price of $480. The additional value to consumers—over and above their expenditures on the service—would be $284 billion to $427 billion per year, assuming that the linear demand curve with a current elasticity of -1.0 or -1.5 simply shifted outward. See Table 11-13 for a side-by-side comparison of these figures.

WiMAX and VoIP   There are a number of distinct economic advantages of WiMAX VoIP over the PSTN and cell phone services. Firstly, there is the decreased cost of cell phone service by using WiMAX VoIP telephony in office or any WiMAX-serviced locale. Second, using WiMAX VoIP in the office can eliminate the cost of long-distance interoffice phone bills. Some 70 percent of corporate telephony is interoffice calling. This is an expense that can be eliminated by moving a company's telephony onto its corporate network. If WiMAX becomes a primary means of access within the company, then WiMAX VoIP would potentially eliminate much of a firm's phone bill.

A firm could eliminate all of its interoffice long-distance expenses by deploying VoIP and WiMAX system. Calls routed over the corporate WAN would free the company from costs associated with long-distance phone service. Local phone service costs could be eliminated

Table 11-13

Estimated Ultimate Annual Consumer Surplus from Increased Broadband Penetration ($ Billions)

	Elasticity of Demand at -1.5	Elasticity of Demand at -1.0
At 8 percent penetration	1.4	2.0
At 50 percent penetration	80	121
At 94 percent penetration	284	427

Source: Crandall and Jackson

as well. If firms employed dual frequency telephone handsets, all interoffice calls could be made on the corporate WAN. Local calls could also be routed to other WiMAX or IP enabled handsets without contact with the PSTN. Other handsets could be reached using the cell phone network.

Soon the demand for broadband will reflect not only the growing potential uses of the Internet but also the prospect for using these broadband connections to obtain voice telephone services currently provided over a narrowband connection. The use of broadband access to carry voice—ordinary telephone calls—as well as data will deliver to consumers substantial savings that are not captured in current demand estimates. Voice communications can be compressed, put in packets, and sent over an IP connection.

The cost savings from integrated access will be significant. Reliable Internet telephony would eliminate the need for second or third lines in households for teenagers or fax machines. The FCC estimated that the average household spent $55 per month on local and long-distance telephone service in 1999, and there were 0.289 additional lines for each household with telephone service.[6]

Within a few years, broadband access will permit consumers to substitute other services for services that now cost $55 per month. The FCC estimates that the average residence spends $34 per month for local telephone service and $21 for long-distance telephone service. Part of that local telephone service cost is for the loop that is used for the broadband service. Consumers continue to incur most of those loop costs when broadband service is used, but they avoid the cost of the analog line card, the voice switch, and the voice transmission lines. Vo802.16 should lower the costs of both local and long-distance telephone service, while providing residences with the equivalent of several telephone lines. Crandall and Jackson estimate

---

[6]FCC, "Trends in Telephone Service," 2nd Report (2000), www.fcc.gov/Bureaus/Common _Carrier/Reports/FCC-State_Link/IAD/trend200.pdf.

[7]The FCC's numbers indicate that the average household with telephone service has 1.289 access lines and pays local service fees of $34 per month. Assuming that all lines cost the same (which is not quite right but is reasonable), the average household with telephone service in 1999 paid $7.62 per month for additional line service. If those households without a second line today place an average value of no more than $3.40 per month for second line of service, then the average household will value a second line at $10 per month or more.

that such savings could average $25 per month per household. In addition, households with broadband service would get the equivalent of multiple voice telephone lines. They estimate that this additional service or option of service could be worth $10 per month to the average household.[7] Thus, in the longer run (say a decade from now), broadband access could deliver voice communications benefits of about $35 per month (or $420 per year) to the average household with telephone service. If we assume that 122.2 million households have telephone service, these benefits would total $51.4 billion per year, assuming no growth in voice usage. The actual value could be much higher.

The substantial economic benefits (principally savings from expenditures on telephone service) created by providing multiple services over a high-speed line almost cover the cost of a high-speed line—we have estimated that benefits of $35 per month are created by a broadband connection that costs $40 per month. These savings are one reason why we believe that it is reasonable to expect that the

Table 11-14

Summary of Consumer Benefits from Universal Broadband Deployment ($ Billions per Year)

Source	Low Estimate	High Estimate
Direct Estimate		
Broadband Access Subscription	284	427
Household Computer/ Network Equipment	13	33
Total Benefits	297	460
Alternative Estimates; Benefits Deriving from		
Shopping	74	257
Entertainment	77	142
Commuting	30	30
Telephone Services	51	51
Telemedicine	40	40
Total Benefits	272	520

fraction of households with high-speed access services will ultimately approach the fraction that has telephone service today. Refer
to Table 11-14 for a list of the benefits of universal broadband deployment.

**Speeding Up the Adoption of Broadband Access Provides
Benefits Earlier**   The present value of the difference between the
base adoption scenario and the much faster adoption scenario of our
previous example is 140 percent of one year's worth of the benefits
of ubiquitous broadband adoption by households.[8] Thus, if one
assumed that broadband, when fully adopted, generated benefits of
$300 billion per year to American consumers, a policy change that
moved our society from the baseline adoption curve to the much
faster curve would generate benefits with a net present value of
about $420 billion.[9] The increase in the present value of producers'
surplus would be about $80 billion. This acceleration is therefore
worth $500 billion to U.S. consumers and producers.

How could speeding up the adoption of a technology have such
massive benefits? The key lies in the substantial benefits that ubiquitous broadband can convey to consumers. Once virtually every-
one has the service, the network effects from developing new
services become very large. Moving these benefits forward a few
years can create very large benefits—even when evaluated from
today's perspective. The powerful advantage of WiMAX over the
current, dominant broadband technologies, DSL and cable modem,
is that in the words of Clayton Christensen, it is "cheaper, simpler,
smaller, and more convenient to use (deploy)." The lack of a requirement for wires and their incumbent, expensive rights-of-way has
the potential to give the wireless service provider a significant
advantage over the wired incumbent.

---

[8]This was calculated using a discount rate of 10 percent and assuming a 2 percent
per year growth in the economy.

[9]These present values are 2.8 and 4.2 times the ultimate value of broadband adoption when evaluated at an interest rate of 10 percent per year.

Conclusion

In a summer 2003 televised interview of Intel founder Andy Grove, interviewer Charlie Rose asked his guest, "What's the next big thing?" Mr. Grove, thinking he was being hit up for the equivalent of a stock tip, fended off his interviewer until Mr. Rose parried with a deeper inquiry mentioning Intel's Centrino chip and other wireless initiatives at Intel. Mr. Grove then jumped in with both feet. "Wireless is the next big thing," he summarized, "It will be bigger than telegraphy or even telephony itself."

At the time of that interview, WiMAX was still an obscure technology, but as Intel is the lead chipmaker offering WiMAX technology, Mr. Grove was no doubt well aware of the potential impact of WiMAX on the telecommunications market and the world as a whole. As pointed out in this chapter, no existing subindustry in telecommunications (cable TV, wire line telephony, cell phones, Internet access) will go untouched.

Projections: WiMAX Is a Disruptive Technology

This is a very exciting time in the telecommunications industry. There is a powerful clamor for service providers to roll out the *triple play* of voice, video, and data. A *quadruple play* may include mobile phone and data services. Given the commonality of IP, all that remains is an inexpensive means of delivering those IP bits to the subscriber, thus banishing the curse of the "last mile bottleneck." WiMAX breaks open that last mile bottleneck.

Disruptive Technology

In his Harvard University business book, *The Innovator's Dilemma*, Clayton Christensen describes how disruptive technologies have precipitated the failure of leading products and their associated and well-managed firms. Christensen defines criteria to identify disruptive technologies, regardless of their market. These technologies have the potential to replace mainstream technologies and their associated products and principal vendors. Christensen abstractly defines disruptive technologies as "typically cheaper, simpler, smaller, and, frequently, more convenient" than their mainstream counterparts.[1]

Wireless technologies, relative to incumbent wired networks, are a disruptive technology. For the competitive service provider, WiMAX *is* "cheaper, simpler, smaller, and, frequently, more convenient" than copper wire or coax cable and their associated infrastructures. In order for a technology to be truly disruptive, it must *disrupt* an incumbent vendor or service provider. Some entity must go out of business before a technology can be considered *disruptive*. Although it is too early to point out incumbent service providers driven out of business by WiMAX, its technologies are potentially disruptive to incumbent telephone companies. The migration of wire line telephone traffic from ILEC to cellular is a powerful example of this trend. The migration to voice over WiMAX will certainly mark the disruption of telephone companies as we know them.

---

[1]Clayton Christensen, *The Innovator's Dilemma* (Boston: Harvard Business School Press, 1997), p. 264.

How WiMAX Will Disrupt the Telephone Industry

If disruptive technology is defined as being "cheaper, simpler, smaller, and more convenient to use," how, then, is WiMAX cheaper, simpler, smaller, and more convenient to use than legacy telecommunications infrastructure?

Cheaper

A WiMAX network is much cheaper to deploy than a comparable TDM-switched, copper wire-based legacy PSTN infrastructure. The Telecommunications Act of 1996 failed to produce any real competition in the local loop, as it was economically impossible to build and deploy a network that could compete with an entrenched and financially protected monopoly.

WiMAX changes all that. As explored earlier, a competitive network can be built for a fraction of the cost of a legacy network. Furthermore, it can be operated for a fraction of the operations, administration, maintenance, and provisioning (OAM&P) of the PSTN. Potentially, it offers more services than the PSTN, generating more revenue than a PSTN voice-centric infrastructure.

By virtue of being cheaper to purchase and operate, a WiMAX network marks a significant lowering of barriers to market entry. No longer is a voice service the exclusive domain of a century-old protected monopoly. This lowering of the barrier to entry will allow multiple types of service providers to offer voice services in direct competition with the legacy telephone monopoly. This list of service providers could include WISPs, ISPs, power companies, municipalities, cable TV companies, and new market entrants. For the price of a new pick-up truck, a rural ISP can be the telephone company, cable TV company, and potentially the cell phone company for a given community. This is very disruptive.

Simpler

Given its 100-year evolution, the PSTN is painfully complex. Service providers have melded one technology on top of another over the last century. COs are, in many cases, museums of switching history, as operators rarely discard switching equipment that still functions (and enjoys a very generous depreciation schedule).

WiMAX service providers will not be burdened by the past. Rather, a WiMAX is IP-based, meaning it is far more efficient in operation. The key here is open standards as opposed to the closed systems of the legacy PSTN. The open standards allow a service provider to mix and match components of the network. Much of a softswitched voice network is software-dependent, which can be upgraded easily and frequently. The use of IP-based media (voice and video) further simplifies service delivery.

Smaller

One recurring excuse for the monopoly of telephone or cable TV companies is that they were/are an "economy of scale," in that something so large, so complex, and so costly could succeed only if it were protected as a monopoly. A WiMAX network can be easily deployed as a modular system by even the smallest service providers in rural or developing economies. The same is true of corporate campuses or multidwelling units (MDUs). Given that VoIP, IPTV, or Internet operations are geographically independent of the subscriber, a service provider can provide switching for widely dispersed subscribers.

The footprint of a WiMAX CPE is comparable to that of a laptop computer. The antenna and radio are also about the size of a laptop. This makes deployment fast and inexpensive. The smaller size makes deployment and management that much easier.

More Convenient to Use

The PSTN may be doomed by voice, the commodity for which it was created. Ditto for cable TV networks and their video equivalent. Business and residential markets now demand a convenient access to broadband data services. The PSTN does not offer this function efficiently. WiMAX networks offer easily deployed and operated broadband data services with the triple play of voice, video, and data. WiMAX works because the flexibility of its all-IP infrastructure offers the subscriber greater convenience (VoIP, IPTV, and IP data from one service provider).

Deconstruction

In their 1999 book titled *Blown to Bits*, Phillip Evans and Thomas Wurster explore how certain industries have been "deconstructed" by the Internet. That is, the emergence of information or services available via the Internet has caused firms to lose sales and market share, if not their entire business, due to the emergence of new technologies. Examples of those industries include travel agencies, retail banks, and automobile retailers. The following pages will investigate the potential deconstruction of the North American telecom industry by Internet-related telephony applications.[2]

The telecom sector in recent years has been deconstructed by technologies that are Internet-related, if not by the Internet itself. The delivery of telephony features to a voice service via IP would also be an example of deconstruction of the telecom service provider industry by an Internet-related technology. IPTV does the same for the cable TV industry. Cell phone companies may find themselves similarly deconstructed, once mobility in WiMAX reaches the market.

Goetterdammerung or Creative Destruction in the Telecommunications Industry

Every month, North American local exchange carriers lose thousands of their TDM line accounts. Furthermore, some are deeply in

---

[2]Phillip Evans and Thomas Wurster, *Blown to Bits* (Boston: Harvard Business School Press, 1999).

debt. Percentage-wise, this marks the only time since the Great Depression that telephone companies have actually decreased in line count.

How could the telephone company lose business? The answer is simple: Competition is slowly coming *to* as opposed to *in* the local loop. Subscribers are taking their business elsewhere. Many competing technologies allow subscribers to divorce themselves from the former monopolies. Many residential subscribers have given all their voice business to their cell phone service provider, and businesses have taken their voice business to data companies that offer VoIP over a data connection (ICG, Vonage). Capital expenditures for telephone companies are at record lows. The near-monopolistic vendors of the past are mired deeply in debt.

Is there no optimism in this market? If one is looking for a recovery in the telecommunications market as we know it, there is no cause for optimism. Austrian-born Harvard economist Josef Schumpeter, if he were alive today, would probably refer to the current telecommunications industry as being a good case of *creative destruction*. That is, capitalism is cyclical: Almost all industries grow, mature, and die.

The telecommunications industry as we know it is no exception to this rule of capitalism. Shielded as a quasi-monopoly for most of its life, the North American local exchange carrier had no reason to compete or to innovate. The service it provides, voice, is little changed over 100 years ago. The monopolistic protection came to an end with the Telecommunications Act of 1996. The resulting boom in the industry buoyed those incumbent carriers as the "high tide that raises all boats." The telecommunications bust has seen the demise of many competitors in the local loop but has yet to seriously threaten the survival of the incumbents. WiMAX may change that. Incumbent telcos that cannot adapt to the challenges posed by WiMAX, VoIP, and IPTV will die.

WiMAX potentially strikes at the very heart of the incumbent telco business paradigm that relied on a high barrier to entry to the voice market. Technology will inevitably march forward. WiMAX technology is "cheaper, simpler, smaller, and more convenient to use." It is a disruptive technology that, after matching the incumbent technology, has qualities of its own that will allow it to supersede the incumbent's legacy infrastructure. WiMAX, unlike incumbent circuit-switched infrastructures, is a technology that can be quickly and cheaply deployed anywhere in the world. The North American telephony market (services) is estimated to do almost $1 trillion in business annually. Service providers, regardless of the technologies they use, will, in a Darwinian struggle, seek to get an ever-increasing larger market share. That market share can come only at the expense of the incumbents.

In summary, there will not be a recovery in the North American telecommunications market. There will be a rebirth. That rebirth will come in the form of new service providers offering new services with new technology. It is not certain when the exact date of the end of circuit-switched telephony and the century-old PSTN will come. The best analogy of this passing is in the Wagnerian opera *"Goetter-daemmerung"* or "twilight of the gods." *"Daemmerung,"* in this case, translates into "twilight," which in the German sense of the word can mean twilight at dusk and at dawn. In the case of the North American telecommunications market, it is the dusk for the incumbents and their legacy voice-centric networks, and it is dawn for WiMAX.

Considerations in Building Wireless Networks

This appendix is not intended as a "how to" guide but rather to give the reader a broad overview of the foibles and tricks of the trade for the installation of wireless networks.

Successful deployment of a WiMAX system requires design, planning, implementation, operation, and maintenance. This chapter provides a very brief overview of what the wireless network planner needs to consider when deploying a WiMAX network.

Design

Some of the questions that must be addressed in selecting a WiMAX solution lead to trade-offs—such as speed versus range. Others have mutually exclusive answers—such as proprietary versus standards-based extensions. Some of the questions that need to be addressed include: What is the network topology? What kinds of links will be used? What is the environment like? What is the throughput, range, and bit error rate (BER) that is needed? Will one need tolerance for multipath? What frequency band will be used with what protocols? Can the solution be off-the-shelf or surplus standards based, or will it need to be custom?

Network Topology

One of the major factors that determine throughput, robustness, reliability, security, and cost is the geometric arrangement of the network components, or the topology. Five major topologies are in use today in wired networks: Bus, Star, Tree, Ring, and Mesh. In Wireless LAN only, the Star and Mesh have analogues with the wired networks.

The mesh topology is a slightly different type of network architecture than the better-known star topology, except that there is no centralized BS. Nodes that are in range of one another can communicate freely, as shown in Figure A-1.

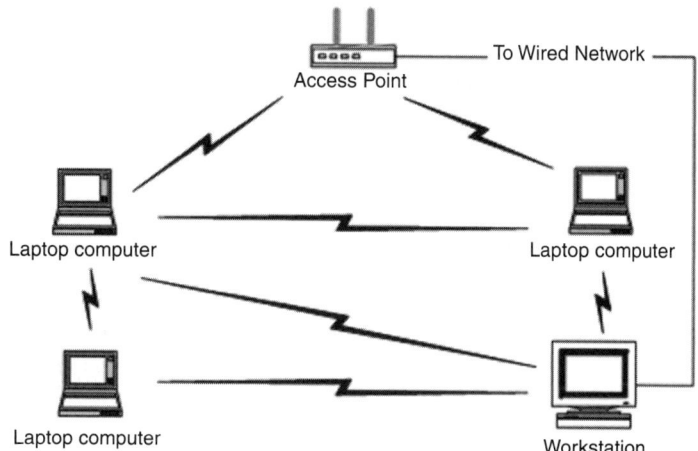

Wireless mesh networks are an exciting new topology for creating low-cost, high-reliability wireless networks in a building, across a campus, or in a metropolitan area. In a mesh network, each wireless node serves as both an AP and wireless router, creating multiple pathways for the wireless signal. Mesh networks have no single point of failure and thus are self-healing. A mesh network can be designed to route around line-of-sight obstacles that can interfere with other wireless network topologies. However, using a wireless mesh currently requires the use of specialized client software that will provide the routing function and put the radio into ad-hoc or infrastructure mode as required.

Link Type

WiMAX systems can be built using either point-to-point or point-to-multipoint links. FCC regulations allow both types of links, but they come with implications for the power to the antenna. See Figure A-2.

Environment

What is the environment like? Is it indoors or outdoors? Is there a line of sight, or are there obstacles in the path? See Figure A-3.

Figure A-2
Point-to-point
and point-to-
multipoint

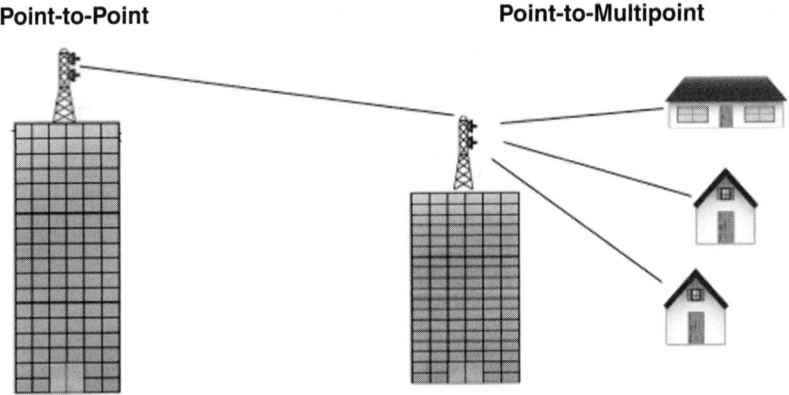

Point-to-Point

Point-to-Multipoint

Figure A-3
Line-of-sight vs.
non-line-of-sight
signal

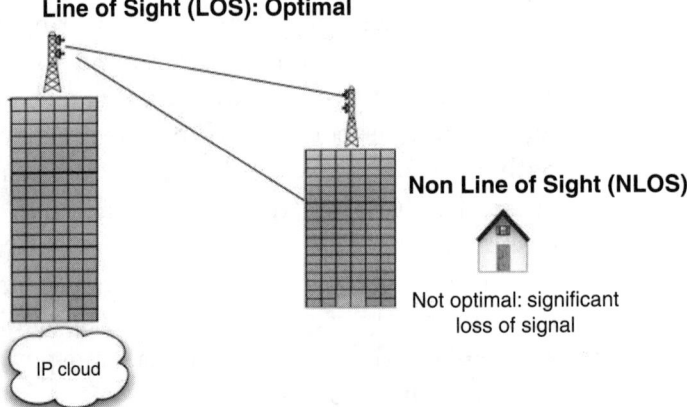

Line of Sight (LOS): Optimal

Non Line of Sight (NLOS)

Not optimal: significant
loss of signal

IP cloud

Throughput, Range, and Bit Error Rate (BER)

Throughput has trade-offs with range and BER. The best network designs balance these factors by limiting the data rate according to data quantity and latency requirements. Fundamentally, in any application there is a trade-off between three factors: range, throughput, and BER. Because the throughput is limited by the protocol (802.11 WiMAX) and because BER has to be reasonably high to get throughput, the only variable left is the range. The available range at a given throughput can be calculated using a link budget.

Multipath Fading Tolerance

Non-line-of-sight positioning must allow for significant multipath fading. Multipath is created by reflections canceling the main signal. The choice of frequency band and protocols will, in part, depend on how much multipath can be tolerated.

Link Budget

A fundamental concept in any communications system is the link budget, or the summation of all the gains and losses in a communications system. The result of the link budget is the transmit power required to present a signal with a given signal-to-noise ratio (SNR) at the receiver to achieve a target BER.

For any wireless protocol, it is sufficient to consider factors such as path loss, noise, receiver sensitivity, and gains and losses from antennas and cable. Before calculating a link budget, factors such as the frequency band must be determined.

Frequency Band

Some wireless technologies can be deployed on four unlicensed frequency bands in two bands called ISM and Unlicensed National Information Infrastructure (U-NII). The 2.4 GHz ISM band has an inherently stronger signal with a longer range and can travel through walls better than the 5 GHz U-NII bands can. However, the U-NII band allows more users to be on the same channel simultaneously. The 2.4 GHz ISM band has a maximum of three non-overlapping 22 MHz channels while the 5 GHz band has four non-overlapping 20 MHz channels in each of the U-NII bands.

Industrial, Scientific, and Medical (ISM) Band   The ISM bands were originally reserved internationally for noncommercial use of RF electromagnetic fields for industrial, scientific, and medical purposes. More recently, they have also been used for license-free error-tolerant communications applications such as cordless phones, Bluetooth, and Wireless LAN.

U-NII Band Devises that will provide short-range, high-speed, wireless, digital communications can use the U-NII bands. These devices, which do not require licensing, will support the creation of wireless metro area networks (WMANs) and facilitate access to the Internet. The U-NII spectrum is located at 5.15 to 5.35 GHz and 5.725 to 5.825 GHz.

The 5.15 to 5.25 GHz portion of the U-NII band is intended for use by indoor, short-range networking devices. The FCC adopted a 200-mW EIRP limit to enable short-range wireless LAN applications in this band without causing interference to mobile satellite service (MSS) feeder link operations.

Devices operating between 5.25 and 5.35 GHz are intended to be communications within and between buildings, such as in campus-type networks. U-NII devices in the 5.25 to 5.35 GHz frequency range are subject to a 1 watt EIRP power limit.

The 5.725 to 5.825 GHz portion of the U-NII band is intended for community networking communications devices operating over longer distances. The FCC permits fixed, point-to-point U-NII devices to operate with up to a 200 Watt EIRP limit.

FCC Regulations The use of these bands is regulated under part 15.247 and 15.407 of the FCC regulations.[1] The following are the relevant parts of part 15.247 regarding power at the time of writing:

(b) The maximum peak output power of the intentional radiator shall not exceed the following:

(1) For frequency hopping systems operating in the 2,400–2,483.5 MHz or 5,725–5,850 MHz band and for all direct sequence systems: 1 watt.

(3) Except as shown in paragraphs (b)(3)(i), (ii) and (iii) of this section, if transmitting antennas of directional gain greater than 6 dBi are used, the peak output power from the intentional radiator shall be reduced below the stated values in paragraphs (b)(1) or (b)(2) of this section, as appropriate, by the amount in dB that the directional gain of the antenna exceeds 6 dBi.

---

[1]The FCC website, www.fcc.gov/, has a lot of material online. Part 15 in its entirety can be found at www.access.gpo.gov/nara/cfr/waisidx_01/47cfr15_01.html.

(i) Systems operating in the 2,400-2,483.5 MHz band that are used exclusively for fixed, point-to-point operations may employ transmitting antennas with directional gain greater than 6 dBi provided the maximum peak output power of the intentional radiator is reduced by 1 dB for every 3 dB that the directional gain of the antenna exceeds 6 dBi.

(ii) Systems operating in the 5,725-5,850 MHz band that are used exclusively for fixed, point-to-point operations may employ transmitting antennas with directional gain greater than 6 dBi without any corresponding reduction in transmitter peak output power.

Part 15.407 regulates the UNII band and its operation. The following parts are the relevant to understanding power limits within the 5.1, 5.2, and 5.8 GHz bands:

(a) Power limits:

(1) For the band 5.15–5.25 GHz, the peak transmit power over the frequency band of operation shall not exceed the lesser of 50 mW or 4 dBm + 10logB, where B is the 26 dB emission bandwidth in MHz. In addition, the peak power spectral density shall not exceed 4 dBm in any 1 MHz band. If transmitting antennas of directional gain greater than 6 dBi are used, both the peak transmit power and the peak power spectral density shall be reduced by the amount in dB that the directional gain of the antenna exceeds 6 dBi.

(2) For the band 5.25–5.35 GHz, the peak transmit power over the frequency band of operation shall not exceed the lesser of 250 mW or 11 dBm + 10logB, where B is the 26 dB emission bandwidth in MHz. In addition, the peak power spectral density shall not exceed 11 dBm in any 1 MHz band. If transmitting antennas of directional gain greater than 6 dBi are used, both the peak transmit power and the peak power spectral density shall be reduced by the amount in dB that the directional gain of the antenna exceeds 6 dBi.

(3) For the band 5.725–5.825 GHz, the peak transmit power over the frequency band of operation shall not exceed the lesser of 1 W or 17 dBm + 10logB, where B is the 26 dB emission bandwidth in MHz. In addition, the peak power

spectral density shall not exceed 17 dBm in any 1 MHz band. If transmitting antennas of directional gain greater than 6 dBi are used, both the peak transmit power and the peak power spectral density shall be reduced by the amount in dB that the directional gain of the antenna exceeds 6 dBi. However, fixed point-to-point U-NII devices operating in this band may employ transmitting antennas with directional gain up to 23 dBi without any corresponding reduction in the transmitter peak output power or peak power spectral density. For fixed, point-to-point U-NII transmitters that employ a directional antenna gain greater than 23 dBi, a 1 dB reduction in peak transmitter power and peak power spectral density for each 1 dB of antenna gain in excess of 23 dBi would be required. Fixed, point-to-point operations exclude the use of point-to-multipoint systems, omni directional applications, and multiple collocated transmitters transmitting the same information. The operator of the U-NII device, or if the equipment is professionally installed, the installer, is responsible for ensuring that systems employing high gain directional antennas are used exclusively for fixed, point-to-point operations.

Table A-1 summarizes the ISM and U-NII unlicensed frequency bands used by WiMAX devices and shows their associated power limits.

Point-to-Multipoint   Part 15.247(b)(1) limits the maximum power at the antenna to 1 watt.

Part 15.247(b)(3) allows antennas that have more than 6 db, as long as the power to the antenna is reduced by an equal amount in the 2.4 GHz band. This implies that the maximum effective isotropic radiated power (EIRP) is 4 watts or 36 dBm.

This limit of 4 watts EIRP irrespective of antenna gain is illustrated in Table A-2.

Point-to-Point Links   Point-to-point links have a single transmitting point and a single receiving point. Typically, a point-to-point link is used in a building-to-building application. Part 15.247 (b)(3)(i) allows the EIRP to increase beyond the 4-watt limit for

Table A-1

Frequency Bands and Associated Power Limits

Frequency Range (MHz)	Bandwidth (MHz)	Max Power at Antenna	Max EIRP	Notes
2,400–2,483.5	83.5	1W (+30dBm)	4W (+36dBm)	Point-to-point
		1W (+30dBm)		Point-to-multipoint following 3:1 rule
5,150–5,250	100	50mW	200mW (+23dBm)	Indoor use; must have integral antenna
5,250–5,350	100	250mW (+24dBm)	1W (+30dBm)	
5,725–5,825	100	1W (+30dBm)	200W (+53dBm)	

Table A-2

Point-to-Multipoint Operation in 2.4 GHz ISM Band

Power at Antenna (mW)	Power at Antenna (dBm)	Max Antenna Gain (dBi)	EIRP (watts)	EIRP (dBm)
1,000	30	6	4	36
500	27	9	4	36
250	24	12	4	36
125	21	16	4	36
63	18	19	4	36
31	15	21	4	36
15	12	24	4	36
8	9	27	4	36
4	6	30	4	36

point-to-multipoint links in the 2.4 GHz ISM band. For every additional 3 db gain on the antenna, the transmitter only needs to be cut back by 1 db.

The so-called three-for-one rule for point-to-point links can be observed in Table A-3.

Table A-3

Point-to-Point Operation in 2.4 GHz ISM band

Power at Antenna (mW)	Power at Antenna (dBm)	Max Antenna Gain (dBi)	EIRP (watts)	EIRP (dBm)
1,000	30	6	4	36
794	29	9	6.3	38
631	28	12	10	40
500	27	15	16	42
398	26	18	25	44
316	25	21	39.8	46
250	24	24	63.1	48
200	23	27	100	50
157	22	30	157	52

Table A-4

Point-to-Point Operation in 5.8 GHz U-NII Band

Power at Antenna (mW)	Power at Antenna (dBm)	Antenna Gain (dBi)	EIRP (watts)	EIRP (dBm)
1,000	30	6	4	36
1,000	30	9	8	39
1,000	30	12	16	42
1,000	30	15	316	45
1,000	30	18	63.1	48
1,000	30	21	125	51
1,000	30	23	250	53

According to part 15.247(b)(3)(ii), there is no such restriction in the 5.8 GHz band. However, part 15.407 effectively restricts the EIRP to 53 dBm. See Table A-4.

Wireless Protocols Preceding WiMAX

Four primary standards-based protocols precede WiMAX: 802.11, 802.11b, 802.11a, and 802.11g.

802.11 The 802.11 standard was the first standard to specify the operation of a WLAN. This standard addresses Frequency Hopping Spread Spectrum (FHSS), Direct Sequence Spread Spectrum (DSSS), and infrared. The data rate is limited to 2 Mb/sec and 1 Mb/sec for both FHSS and DSSS.

FHSS handles multipath and narrowband interference as well as a by-product of its frequency-hopping scheme. If multipath fades one channel, other channels are usually not faded. Thus, packets are passed on those hops where no fading occurs. Operating an FHSS system in a high-multipath or high-noise environment will be seen as an increase in latency. FHSS has 64 hopping patterns, which can support up to 15 collocated networks. FHSS systems are limited to 1 Mb/sec and optionally 2+ Mb/sec. Typically, they have a shorter range than DSSS systems have. FHSS is not compatible with today's 802.11b equipment.

DSSS as implemented in 802.11 occupies 22 MHz of spectrum while providing a maximum over-the-air data rate of 2 Mb/sec. DSSS is susceptible to multipath and narrowband interference due to the limited amount of spreading that is used (11 bits). DSSS can only support three noninterfering channels and thus does not have nearly as much network capacity as an FHSS system at the same data rate. DSSS is compatible with today's 802.11b equipment.

Surplus 802.11 equipment may work well for some applications where multipath immunity is required, lower data rates can be tolerated, and compatibility with currently available equipment is not desired. Furthermore, be advised that the gear may no longer be covered by warranties and may not have service available for it anymore.

802.11b   The most widely used standard protocol, 802.11b, requires DSSS technology, specifying a maximum over-the-air data rate of 11 Mb/sec and a scheme to reduce the data rate when higher data rates cannot be sustained. This protocol supports 5.5 Mb/sec, 2 Mb/sec, and 1 Mb/sec over-the-air data rates in addition to 11 Mb/sec using DSSS and CCK.

IEEE 802.11b standard uses complementary code keying (CCK) as the modulation scheme to achieve data rates of 5 Mb/sec and 11 Mb/sec. 802.11 reduced the spreading from 11 bits to 8 to achieve the higher data rates. The modulation scheme makes up the processing gain lost with the lower spreading by using more forward error correction (FEC).

The IEEE 802.11b specification allows for the wireless transmission of approximately 11 Mbps of raw data at indoor distances to about 300 feet and outdoor distances of perhaps 20 miles in a point-to-point use of the 2.4 GHz band. The distance depends on impediments, materials, and line-of-sight.

802.11b is the most commonly deployed standard in public short-range networks, such as those found at airports, coffee shops, hotels, conference centers, restaurants, bookstores, and other locations. Several carriers currently offer pay-as-you-go hourly, per session, or unlimited monthly access, using networks in many locations around the United States and other countries.

802.11a   The 802.11a standard operates in the three 5 GHz U-NII bands and thus is not compatible with 802.11b. The bands are designated by application. The 5.1 GHz band is specified for indoor use only, the 5.2 GHz band is designated for indoor/outdoor use, and the 5.7 GHz band is designated for outdoor only. RF interference is much less likely because of the less-crowded 5 GHz bands. The 5 GHz bands each have four separate nonoverlapping channels. It specifies OFDM using 52 subcarriers for interference and multipath avoidance; supports a maximum data rate of 54 Mb/sec using 64-QAM; and mandates support of 6, 12, and 24 Mb/sec data rates. The protocol specifies minimum receive sensitivities ranging from −65 dBm for the 54 Mb/sec rate to −82 dBm for 6 Mb/sec. Equipment designed for the 5.1 GHz band has an integrated antenna and is not easily modified for higher power output and operation on the other two 5 GHz bands.

802.11g   802.11g is an extension to 802.11b and operates in the 2.4 GHz band. 802.11g increases 802.11b's data rates to 54 Mbps using the same orthogonal frequency division multiplexing (OFDM) technology that is used in 802.11a. The range at 54 Mbps is less than existing 802.11b APs operating at 11 Mbps. As a result, if an 802.11b cell is upgraded to 802.11b, the high data rates will not be available throughout all areas. You'll probably need to add additional APs and replan the RF frequencies to split the existing cells into smaller ones. 802.11g offers higher data rates and more multipath tolerance. Although there is more interference on the 2.4 GHz band, 802.11g may the protocol of choice for best range and bandwidth combination, and it's upwardly compatible with 802.11b equipment.

802.11 Summary

Which technique is best? It depends on the application and other design considerations. Frequency hopping offers superior reliability in noise and multipath fading environments. Direct sequence can provide higher over-the-air data rates. OFDM offers multipath tolerance and much higher data rates. 802.11 (no letter) is now obsolete but may offer nonstandard, bargain-basement, usable equipment. 802.11b is compatible with most of the public access locations. 802.11a is the best to solve interference cases and has great throughput. 802.11g promises the best range and throughput combination of all the solutions.

Planning

A link budget will tell what is practical given the environment and how to plan cells. With a link budget, one can estimate how many cells will be required for the project. There are trade-offs between more cells and running more power. For an outdoor application, also consider checking the Fresnel zone. For example, the trade-off between working on one long-distance shot versus two back-to-back links can be discovered by working out a few things on paper first.

Fresnel Zone

For a point-to-point shot, it is important to understand the effects of the Fresnel zone. The signal extending from the transmitting antenna forms an ever-expanding cone. Although some of the signal travels directly in the line of sight along the center of the cone, other parts of the signal reradiate off of points along the way. At the receiver, signals from the direct line of sight indirectly cancel and add to each other. The first Fresnel zone is the surface containing every point for which the sum of the distances from that point to the two ends of the path is exactly ½ wavelength longer than the direct end-to-end path. The second Fresnel zone is the surface containing every point for which the sum of the distances from that point to the two ends of the path is exactly one wavelength longer than the direct end-to-end path. Figure A-4 shows the first and second Fresnel zones.

One can calculate the perpendicular distance of the Fresnel zone from the line that connects the transmitter and the receiver using the following Nth Fresnel zone formula:

$$F_N = \sqrt{\frac{N \times \lambda \times D_1 \times D_2}{D_1 + D_2}}$$

Figure A-4
First and second
Fresnel zones

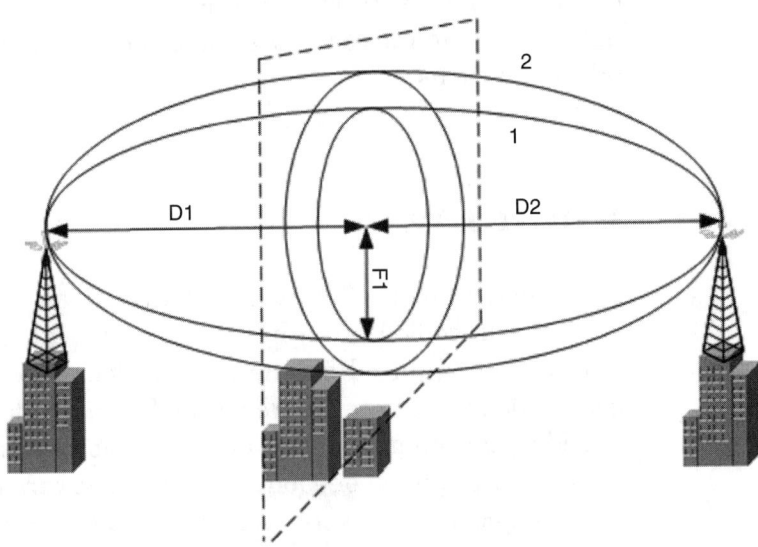

Where:

N is the Fresnel zone number, N = 1 is the first Fresnel zone

λ is the wavelength [meters]

D1 and D2 are the distances to the endpoints [meters]

If there are reflections of the signal from the odd number Fresnel zone, the signal level will cancel at the receiver, but if the reflection is from an even number Fresnel zone, it will add at the receiver. Therefore, on long-distance shots, it is necessary to take into account ground/water reflections and vertical surfaces such as tall buildings.

Because the majority of the transmitted power is in the first Fresnel zone, any time the path clearance between the terrain and the line-of-sight path is less than $0.6F_1$ (six-tenths of the first Fresnel zone distance), some knife-edge diffraction loss will occur. The amount of loss depends on the amount of penetration. To find out if there is any building or obstruction in the Fresnel zone, a profile of the terrain is superimposed with the ellipse created by the Nth Fresnel zone formula with the first Fresnel zone (N = 1) and the result is multiplied by 0.6 for repetitive points across the profile.

Signal strength is possibly gained at the receiver up to 3 dB by having a flat surface such as a lake, a highway, or a smooth desert area at the second Fresnel zone in such a way that the signals get re-enforced at the receiver.

Decibels and Signal Strength   Rather than tracking all those zeroes, amplifier power is measured in a logarithmic scale—decibels (dB). Then instead of multiplying and dividing all the gains, it's much simpler to add and subtract dBs.

$$dB = 10 \times log10 \text{ (power out/power in)}$$

Decibel readings are positive when the output is larger than the input and negative when the output is smaller than the input. Each 10 dB change corresponds to a factor of 10, and 3 dB changes are a factor of 2. Thus, a 33 dB change corresponds to a factor of | 2,000:

$$33dB = 10dB + 10dB + 10dB + 3dB = 10 \times 10 \times 10 \times 2 = 2,000$$

Power is sometimes measured in dBm, which stands for dB above one milliwatt. To find the dBm ratio, simply use 1 mW as the input in the first equation. It's helpful to remember that doubling the power is a 3 dB increase. A 1 dB increase is roughly equivalent to a power increase of 1.25. And a 10 dB increase in power is a 10 times power increase. With these numbers in mind, you can quickly perform most gain calculations in your head.

How to Calculate a Link Budget

Link budget planning is an essential part of the network planning process for both indoor and outdoor applications. A link budget helps to give dimension to the required coverage, capacity, and quality of service requirement in the network. In a typical WiMAX link, there are two link budget calculations: one link from the BS to the SS and the other link from the SS to the BS. Link budgets can be used to determine if the link design meets the designer's criteria for range, throughput, and BER.

A link budget basically adds all the gains and losses to the transmitter power (in dB) to yield the received power. In order to have adequate signal at the receiver, the power presented to the receiver must be at least as much as the receive sensitivity. The link budget for the AP to the client adapter card is shown in Figure A-5.

Transmit Power

Transmitter output power—legal limits

Transmit antenna gain—legal limits

Figure A-5
Calculating a
link budget

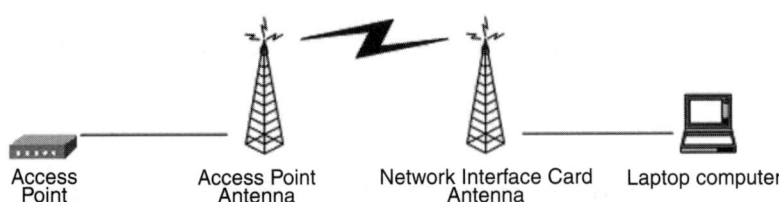

| Access Point | Access Point Antenna | Network Interface Card Antenna | Laptop computer |

Path Loss The most difficult part of calculating a link budget is the path loss. Outdoors, the free-space loss is well understood. The path loss equation[2] for outdoors can be expressed as:

$$\text{Free Space Path Loss} = 20 \log (d \text{ [meters]}) + 20 \log (f \text{ [MHz]}) + 36.6 \text{ dB}$$

At 2.4 GHz, the formula simplifies to:

$$\text{Free Space Path Loss} = 20 \log (d \text{ [meters]}) + 40 \text{ dB}$$

This formula holds true as long as one can see along the line-of-sight from the receiver to the transmitter and have a sufficient amount of area around that path called the Fresnel zone. For indoors, this formula is more complicated and depends on factors such as building materials, furniture, and occupants.

At 2.4 GHz, one estimate follows this formula:

$$\text{Indoor Path Loss (2.4 GHz)} = 55 \text{ dB} + 0.3 \text{ dB} / d \text{ [meters]}$$

At 5.7 GHz, the formula looks like this:

$$\text{Indoor Path Loss (5.7 GHz)} = 63 \text{ dB} + 0.3 \text{ dB} / d \text{ [meters]}$$

Receive Antenna Gain The receive antenna, like the transmit antenna, adds gain into the link budget. Adding gain to an antenna is balanced gain because it adds gain for both transmit and receive.

Link Margin Fade margin is the difference, in dB, between the magnitude of the received signal at the receiver input and the minimum level of signal determined for reliable operation. Links with higher fade margins are more reliable. The exact amount of fade margin required depends on the desired reliability of the link, but a good rule-of-thumb is 20 to 30 dB

---

[2]Edward C. Jordan, *Reference Data for Engineers: Radio, Electronics, Computer, and Communications* (Indianapolis: Howard W. Sams and Co., 1986).

Fade margin is often referred to as "thermal" or "system operating margin."

Diffraction Losses Diffraction occurs when the radio path between the transmitter and receiver is obstructed by a surface that has sharp irregularities or edge. The secondary waves resulting from the obstructing surface are present behind the obstacle. On close to line of sight, diffraction losses can be as little as 6 dB. On non-line-of-sight obstacles, diffraction losses can be 20 to 40 dB.

Coax and Connector Losses Connector losses can be estimated at 0.5 dB per connection. Cable losses are a function of cable type, thickness, and length. Generally speaking, thicker and better-built cables have lower losses (and higher costs). As can be seen from Table A-5, coax losses are nearly prohibitive in the 2.4 and 5.8 GHz bands. The best option is to minimize cable loss and locate the microwave transceiver as close to the antenna as possible in an environmental enclosure.

Attenuation Earlier wireless technologies have been hampered by rain and fog. Wi-Fi and WiMAX are considerably better at dealing with rain fades and other atmospheric degradations.

Rain and Fog When deploying in a rainy or foggy climate, it may be necessary to plan for additional signal loss due to rain or fog. For example, 2.4 GHz signals may be attenuated by up to 0.05 dB/km (0.08 dB/mile) by heavy rain (4 inches/hr). Thick fog produces up to 0.02 dB/km (0.03 dB/mile) attenuation. At 5.8 GHz, heavy rain may produce up to 0.5 dB/km (0.8 dB/mile) attenuation, and thick fog may produce up to 0.07 dB/km (0.11 dB/mile). Even though rain itself does not cause major propagation problems, rain will collect on the leaves of trees and will produce attenuation until it evaporates.

Table A-5

Coax
Attenuation
Losses

Cable Type	2.4 GHz		5.8 GHz	
	dB/100 ft	dB/100 m	dB/100 ft	dB/100m
RG-58	32.2	105.6	51.6	169.2
RG-8X	23.1	75.8	40.9	134.2
LMR-240	12.9	42.3	20.4	66.9
RG^{213}/$_{214}$	15.2	49.9	28.6	93.8
9913	7.7	25.3	13.8	45.3
LMR-400	6.8	22.3	10.8	35.4
3/8 " LDF	5.9	19.4	8.1	26.6
LMR-600	4.4	14.4	7.3	23.9
1/2 " LDF	3.9	12.8	6.6	21.6
7/8 " LDF	2.3	7.5	3.8	12.5
1 1/4 " LDF	1.7	5.6	2.8	9.2
1 5/8 " LDF	1.4	4.6	2.5	8.2

(Source: Times Microwave, Andrew and Belden)

Trees Trees can be a significant source of path loss.[3] A number of variables are involved: What specific type is the tree? Is it wet or dry? And, if it's a deciduous tree, are the leaves present or not? Isolated trees are not usually a major problem, but a dense forest is another story. The attenuation depends on the distance the signal must penetrate through the forest, and it increases with frequency. The attenuation is of the order of 0.35 dB/m at 2.4. This adds up to a lot of path loss if a signal must penetrate several hundred meters of forest!

Fiberglass The loss for a radome is about .5 to 1.0 dB.

[3] "A Generic Vegetation Attenuation Model 1–60 GHz: PM3035," www.radio.gov.uk/topics/research/topics/propagation/vegetation/veg-attenuation-model.pdf.

Glass A normal, clear glass pane will lose about 3 dB at 2.4 GHz. Although most glass will not affect radio frequency, certain kinds of glass severely attenuate signals.[4] It depends on the glass and the tint material. If the glass is at all reflective on either side, chances are that a signal may not be able to penetrate it. New construction often uses tinted, coated, or High-E glass that is designed to hold heat out, and this glass attenuates 802.11 and 802.16 signals. Although High-E glass is not necessarily tinted, it is energy-efficient and is usually double paned, coated, and filled with argon or other inert gasses. Tin oxide (SnO_2) coatings do not pass RF. Some windows have as much as 20 dB loss. Note: Removing external tinting with a razor blade may allow RF to pass through the glass.

Other Building Materials Examples of attenuation through various building materials are shown in Table A-6.[5]

Table A-6

Attenuation of Various Building Materials

Material	Attenuation
Window Brick Wall	2 dB
Metal Frame Glass Wall into Building	6 dB
Office Wall	6 dB
Metal Door in Office Wall	6 dB
Cinder Block Wall	4 dB
Metal Door in Brick Wall	12.4 dB
Brick Wall next to Metal Door	3 dB

[4]"Glass That Cuts Signals," www.isp-planet.com/fixed_wireless/technology/2001/tint_bol.html.

[5]John C. Stein, "Indoor Radio WLAN Performance Part II Range Performance in a Dense Office Environment," Intersil Corp., 1997, available online at http://whitepapers.silicon.com/0,39024759,60016337p-39000370q,00.htm.

Examples

A company claims that a distance of 4.3 miles or 7Km can be spanned in a point-to-multipoint application with their antenna using WiMAX. (Assume no coax losses, no connector losses, and perfect line-of-sight.) Does it work with a minimum spec card? Does it work with the best card?

> 36 dBW 4 W EIRP (max power out point-to-multipoint; includes power out and ant gain)
>
> -116.9 dB (Path loss for 7Km is $40 + 20 \log(7000m)$)
>
> $+ 2.2$ dBi (antenna gain of client adapter card)
>
> -20 dB (link margin)
>
> $= 98.7$ dBm (minimum received power)

The worst-case receive sensitivity for a WiMAX SS is -80 dBm at 1 Mbps. Quite clearly, the signal is not adequate. And this is only barely enough to leave a link margin of 1.3 dB, which is not a very reliable link. Assume its specs are 1 Mbps: -94 dBm; 2 Mbps: -91 dBm; 5.5 Mbps: -89 dBm; 11 Mbps: -85 dBm; and will yield a link margin of 15.3 dB using a 1 Mbps link.

But can you get back at the base station?

> 20 dBm (max output of SS)
>
> $+ 2.2$ dBi (antenna gain of client adapter card)
>
> -116.8 dBm (path loss for 7 Km at 2.4 GHz is $40 + 20 \log(7000m)$)
>
> $- 20$ dB (link margin)
>
> $+ 18$ dB gain (best guess at the gain of a 2ft $\times$ 2ft phased antenna)
>
> $= 96.6 -$ dBm (minimum received power)

If the link margin is 15 dB or the antenna has 23 dB gain, the SS can be heard at 7Km with this antenna.

Given current technology, what is the best you can do on a point-to-point link?

- Get more transmit power and better receive sensitivity.

- Remove noise.

- Limit any attenuation from the link budget.

- Get antennas with the most gain for both ends.

- Go to a lower data rate—better sensitivity (higher data rate = less power efficiency).

- Use two antennas for diversity at both ends.

Assume an SS to have output power at 100 mW or 20 dBm and a very sensitive receiver at −85 dBm at 11 mbps.

If you now go to a lower data rate, for example 1 Mbps, the receive sensitivity is at −94 dBm. A link budget for a record-breaking point-to-point link looks like this:

24 dBm (max legal output of transmitter)

+ 24 dBi (grid antenna gain)

− 20 dB (link margin)

− 2 dB (connector losses)

− 2 dB (coax losses)

+ 24 dB gain (grid antenna gain)

+ 94 dBm (minimum received power at 1 mbps)

= maximum path loss = 142 dbm

or about 75 miles [path loss for 120 Km at 2.4 GHz is 40 + 20 log(120,000m)]

Under full multipath conditions, this link will have a 1 megabit data rate. Under better conditions, the link may operate at the full data rate of 11 Mbps.

Site Survey

Once things work on paper with an adequate link budget and the Fresnel zone, one can go out to the site and see if the paper plan works.

Outdoor Site Survey All the data on paper may indicate that everything will work for a particular link—the link budget, the Fresnel zone can be checked against a topographical view of the point-to-point shot. You may even have used expensive ray-tracing programs to predict the path, but there is only one way to learn if the installation will work.

To perform an outdoor site survey for a point-to-point shot, take along binoculars, two-way radios or cell phones, topographical maps, a GPS, a spectrum analyzer, an inexpensive dB attenuator, and radio equipment to take the trial shot. Have a friend go to the other hill and talk to you on the radio.

Drive out to the proposed site and see with the binoculars if the shot is clear. Check for trees or buildings that may have grown up in the path. The best time to plan a long-distance shot is during spring when everything is wet and growing. If it's the dead of winter, remember that the trees will soon grow leaves again.

Using the location of both endpoints, calculate a bearing and tilt angle to point the antenna. Most GPSs have a built-in function to do this. High-gain dishes are more difficult to aim the further out you go. Take both dishes and point them roughly toward each other. Transmit a signal into one dish. With WiMAX gear, link test software allows you to send a series of management frames.

You can take the output of the other dish and feed it into the spectrum analyzer. You will see a display of frequency (across) versus amplitude (up and down). Pick the channel that has the least amount of noise.

Once you get the antennas close, you will see a spike on the frequency the transmitting dish is tuned to; this spike will be surrounded by noise. Sweep the antenna on each end one at a time, and lock-down the antenna at the point where the signal is the strongest.

At this point, you should have sufficient signal-to-noise ratio to receive the signal with a sufficient margin.

The dB attenuators can be used inline to check to see if the link margin is adequate. With 15 dB of attenuation inline, a link should last easily for a few hours. If not, you need to plan on larger dishes and amplifiers.

How to Make a Frequency Plan

After completing an RF site survey, you'll have a good idea of the number and location of APs necessary to provide adequate coverage and performance for users.

Sample Frequency Plan: 2.4 GHz Frequency Reuse   The 2.4 GHz band has eleven 22 MHz-wide channels defined every 5 MHz, going from 2.412 GHz to 2.462 GHz. The 2.4 GHz band has three nonoverlapping channels (1, 6, and 11), as shown in Figure A-6.

These nonoverlapping channels can be used in a three-to-one reuse pattern, as shown in Figure A-7.

Another Example: 5 GHz Frequency Reuse   The operating channel center frequencies are defined at every integral multiple of 5 MHz above 5 GHz. The valid operating channel numbers are 36, 40, 44, 48, 52, 56, 60, 64, 149, 153, 157, and 161. The lower and middle U-NII subbands accommodate eight channels in a total bandwidth of 200 MHz. The upper U-NII band accommodates four channels in a 100 MHz bandwidth. The centers of the outermost channels

Figure A-6
2.4 GHz has three non-overlapping channels.

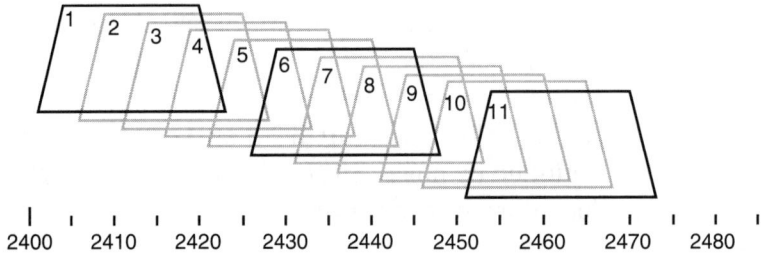

Figure A-7
Three-to-one
reuse pattern

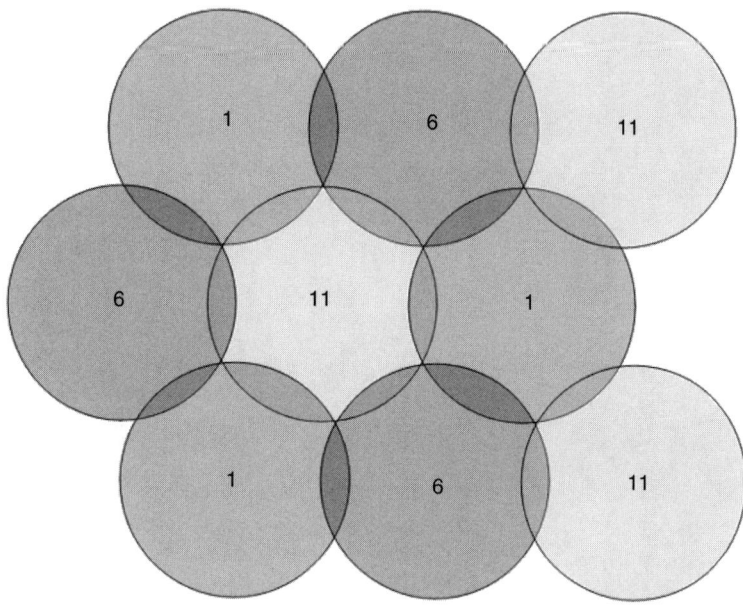

Figure A-8
5 GHz channels

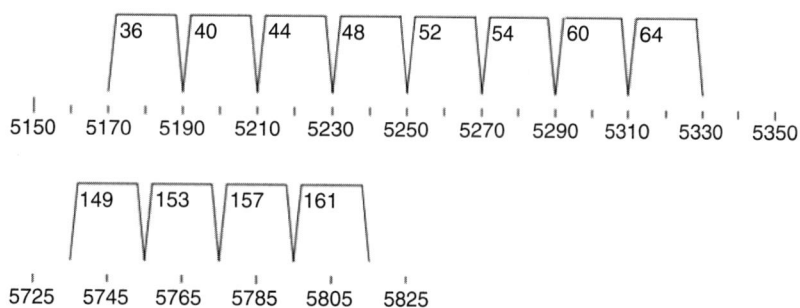

are 30 MHz from the bands' edges for the lower and middle U-NII bands and 20 MHz for the upper U-NII band (see Figure A-8). Point-to-point links operate on the other four channels: 149, 153, 157, and 161. This allows four channels to be used in the same area.

802.11a APs and client adapter cards operate on eight channels: 36, 40, 44, 48, 52, 56, 60, and 64. This allows two four-to-one reuse

Figure A-9
Four-to-one
reuse pattern

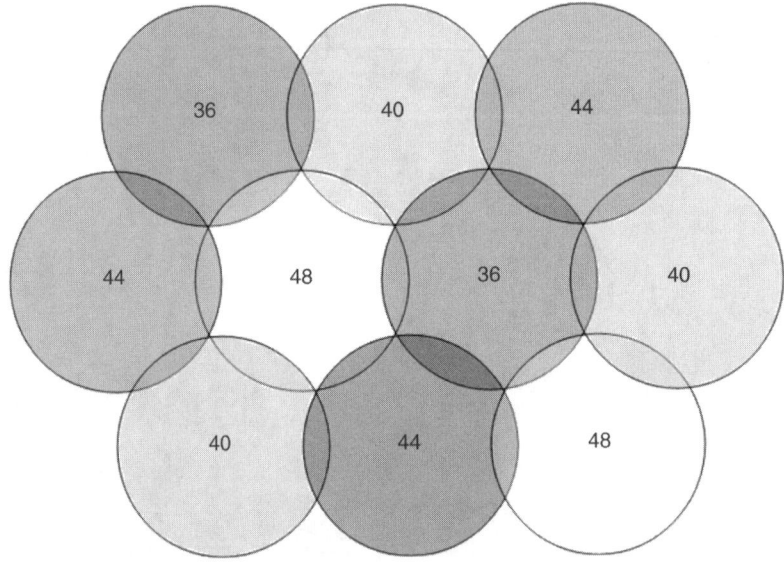

patterns to be used (see Figure A-9). Using both the low- and mid-frequency ranges together allows a seven-to-one reuse pattern with a spare. The spare can be added for a fill to extend coverage or to add capacity in areas such as conference rooms where more capacity is needed (see Figure A-10).

Frequency Allocation

For a simple project such as one or two BSs, simply assign the least used frequencies from the site survey.

For more complex projects involving three or more BSs, pick a frequency reuse pattern for the frequencies that are used for the project, start with the most complicated part of your site survey, and start assigning frequencies. Initially, plan the location of APs for coverage, not capacity. Avoid overlapping channels, if possible. However, if an area has to be overlapped, plan it so that it is naturally an area where the most capacity would be required, such as in a library, conference room, or lecture hall.

Figure A-10
Seven-to-one
reuse pattern
with a spare

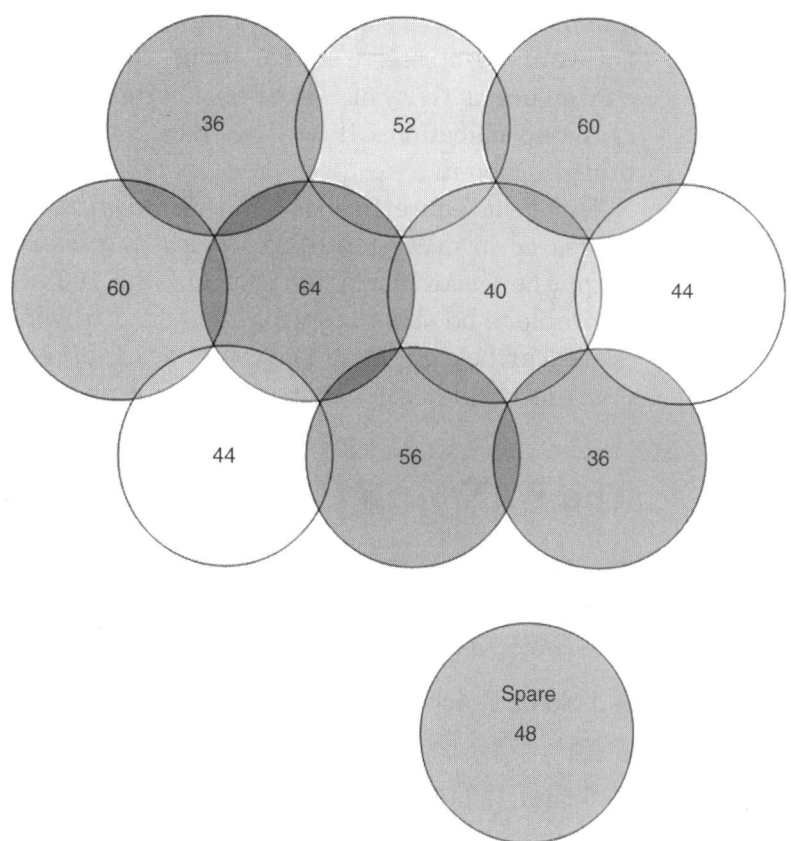

Equipment Selection

The following paragraphs do not necessarily constitute a buyer's guide, but rather a guide to reading a vendor's spec sheet for radios and antennas.

How to Look at Specs

Perhaps the most important spec to consider when looking for wireless equipment is *receive sensitivity*. This is signal strength required for the card to overcome channel noise. Better receiving (a lower dB number) sensitivity means less signal is needed to acquire a signal. For example, a receive sensitivity of −86 dBm may be all right, but a receive sensitivity of −91 dBm is better. Usually this figure is part of the specifications. If it is not listed, then usually it's not worth bragging about.

The next figure to look for is *transmitter output*. This spec is expressed in mW or in dBm. Typically, a transmitter will have an output between 20 mW (or 13 dBm) and 100 mW (or 20 dBm). It is desirable to be able to control the output power so that interference issues can be mitigated. The combination of receiver sensitivity and transmitter are major contributors to range.

The WAN/MAN Connection

The Internet backbone and the WiMAX BSs have plenty of bandwidth, but the WAN connection to the Internet is bandwidth-limited. The choices currently available are as follows:

- DS3 or Fractional DS3
- T1
- Frame Relay
- Cable
- DSL
- ISDN
- Wireless

Basically, you get what you pay for. DS3 or Fractional DS3, T1, and Frame Relay are point-to-point services that are also provided by the LECs and don't come with Internet service. To get Internet service, you need to run the backhaul to an ISP, collocation facility/data center, a "lit" building or to a network access point and become your own ISP. QoS is maintained throughout the network. Of course, T1 and Frame Relay are priced as a business and thus are available at a much higher cost. The bandwidth of a T1 or Frame Relay is 1.5

Mbps. Fractional DS3 is an aggregate of several T1s. Bandwidth is multiples of 1.5 Mbps up to 45 Mbps. It's priced accordingly. The more bandwidth one contracts to buy, the lower the price per 1.5 Mbps increments. Some vendors will also sell by the Mbps.

Antennas Antennas offer another way to increase the range. Antennas limit energy directed in certain areas and redirect the energy in other areas. All antennas exhibit this to a certain extent. A theoretical antenna point source called isotropic is used as a reference for all other antennas. Thus, the gain of an antenna is measured in terms of dBi or decibels over isotropic. Omnidirectional antennas generally have between 2 and 10 dBi, whereas directional antennas can have between 3 and 25 dBi of gain.

FCC regulations limit how much gain a transmitting antenna can have. But antennas have two distinct advantages over amplifiers. First, an antenna offers gains in both the transmit side and the receive side. Thus, the impact on the link budget is balanced. Second, antennas help the interference problem. The transmitter only transmits the signal where it is needed, and the receiver only listens where the antenna is pointed. Not transmitting where other users are and receiving more of the intended signal and less of the interfering station (unless, of course, the interfering station is located in the same antenna path as the intended station) limits interference.

Antennas—BS Side It all starts at the base station. The base station antenna is not the place to economize.

Omnidirectional Omnidirectional antennas transmit their signal roughly equally in all horizontal directions. The radiation pattern has the shape of a large donut around the vertical axis as in Figure A-11.

The gain is in the horizontal direction at the expense of coverage above and below the antenna. For more gain or an outdoor omnidirectional antenna, consider a collinear antenna. Typically, a collinear omnidirectional antenna looks like a PVC pipe that is between 1 and 5 feet tall and has an N connector at the end (see Figure A-12). Gain for these antennas is between 3 and 12 dBi.

Figure A-11
Coverage from
an
omnidirectional
antenna

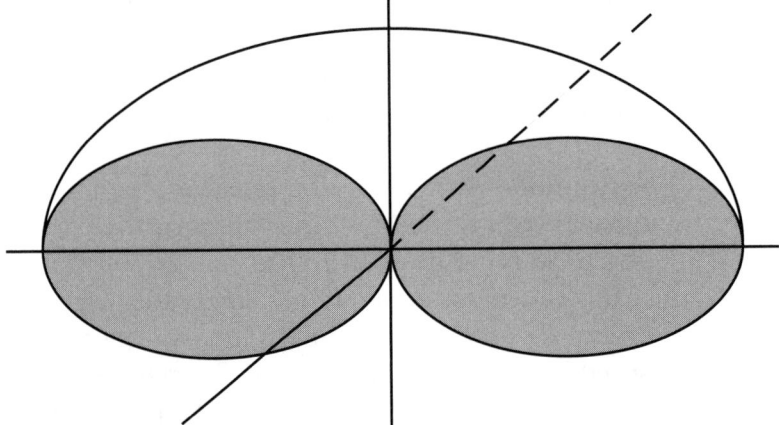

Figure A-12
Collinear
antenna

Vertical   This is a garden-variety omnidirectional antenna. Most vendors sell several different types of vertical antennas, differing primarily in their gain; you might see a vertical antenna with a published gain as high as 10 dBi or as low as 3 dBi. How does an omnidirectional antenna generate gain? Remember that a vertical antenna is omnidirectional only in the horizontal plane. In three dimensions, its radiation pattern looks like a donut. A higher gain means that the donut is squashed. It also means that the antenna is larger and more expensive.

Vertical antennas are good at radiating out horizontally; they're not good at radiating up or down. In a situation like this, it is better to mount the antenna outside a first- or second-story window.

Dipole A dipole antenna has a figure-eight radiation pattern, which means it's ideal for covering a long, thin area. Physically, it won't look much different than a vertical antenna, and some vertical antennas are simply vertically mounted dipoles.

Directional The coverage pattern for a directional antenna looks like Figure A-13. The gain for a patch antenna is typically between 3 and 15 dBi and has a wide beam width.

Sector panel antennas are often used outdoors to cover a sector of a cell. They typically cover 180, 120, or 90 degrees in beam width and have gains between 12 and 20 dBi. These antennas are commonly fitted with an N connector. A panel antenna is shown is Figure A-14.

Yagi For a point-to-point shot, consider a Yagi antenna. A Yagi antenna is a moderately high-gain unidirectional antenna. It resembles a classic TV antenna or washers threaded on a rod. Yagi antennas are often mounted inside of PVC piping to protect them from the weather. There are a number of parallel metal elements at right angles to a boom. Commercially-made Yagis are enclosed in a radome, a plastic shell that protects the antenna from the elements in outdoor deployments. Aiming them is not as difficult as aiming a parabolic antenna though it can be tricky. A Yagi in a radome can be seen in Figure A-15. The beam width and gain is fairly high, 15–20 dBi.

Figure A-13

Coverage from a directional antenna

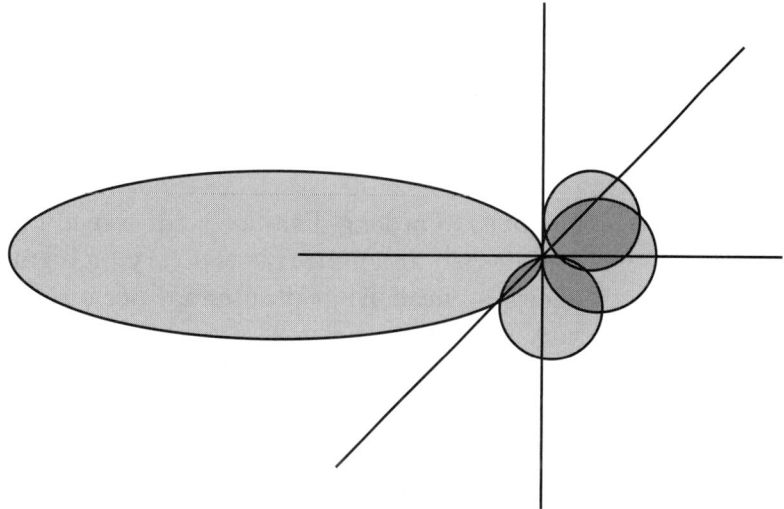

Figure A-14
Panel antenna

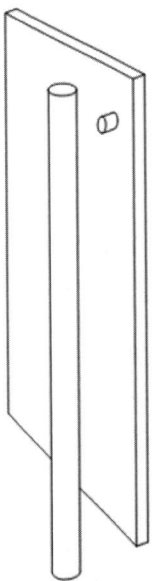

Figure A-15
Yagi antenna in
a radome

Parabolic For long-distance point-to-point shots, choose a parabolic grid or dish antenna. This is a very high-gain antenna. Figure A-16 shows a parabolic grid antenna. Because parabolic antennas have very high gains (up to 24 dBi for commercially made 802.11 antennas), they also have very narrow beam widths. Parabolic antennas are used for links between buildings. Because of the narrow beam width, they are not useful for providing services to end users. Vendors publish ranges of up to 20 miles for their parabolic antennas.

Figure A-16
Parabolic grid
antenna

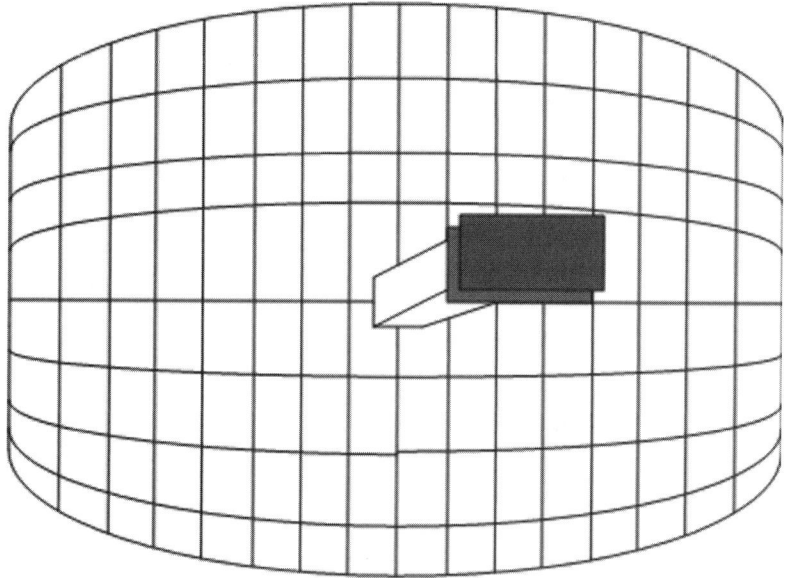

Presumably, both ends of the link use a similar antenna. Front-to-back ratios and wind load are important factors to consider in parabolic grid antennas.

Antenna Specifications   Table A-7 shows typical specifications for antennas and how to interpret them.

Gain   The gain of the antenna is the extent to which it enhances the signal in its preferred direction. Antenna gain is measured in dBi, which stands for decibels relative to an isotropic radiator. An isotropic radiator theoretically radiates equally in all directions. Simple external antennas typically have gains of 3 to 7 dBi. Directional antennas can have gains as high as 24 dBi.

Half-Power Beam Width   This is the width of the antenna's radiation pattern, measured in terms of the points at which the antenna's radiation drops to half of its peak value. Understanding the half-power beam width is important to understanding your antenna's effective coverage area. For a very high-gain antenna, the half-power beam width may be only a couple of degrees. Once outside the half-power beam width, the signal typically drops off quickly, depending

Table A-7

Antenna
Specifications

Specification Name	Description
Frequency Range	Should cover at least 2.4 – 2.4835 MHz.
Gain	Should be expressed in dBi. This figure depends on the antenna.
VSWR	1.5:1 or 2:1 Max typical; lower is better.
Polarization	Vertical, horizontal, or circular.
Half-Power Beam Width	Degrees for vertical and horizontal. This depends on the purpose of the antenna.
Front-to-Back Ratio	This depends on the purpose of the antenna.
Power Handling	Should handle the transmitter's output power $\times$ 3.
Impedance	Should match the transmitter. Usually 50 Ohms.
Connector	N-female is common because it is the strongest, but others are available on commonly available antennas.

on the antenna's design. An antenna's receiving properties are identical to its transmitting properties. An antenna enhances a received signal to the same extent that it enhances the transmitted signal.

Nonstandard Connectors Unlicensed transmitters operating under Section 15.203 are required to be designed so that no antenna other than the one furnished by the party responsible for certifying compliance is used with the device. This can be accomplished by using a permanently attached antenna or a unique coupling at the antenna and at any cable connector between the transmitter and the antenna. FCC Part 15.203 states that intentional radiators operating under this rule shall be designed so that no antenna other than that furnished with it by the responsible party shall be used with the device. The reason for adopting this rule was to prevent the use of unapproved, aftermarket high-gain antennas or third-party amplifiers with a device or system.

To meet this requirement, FCC allows several options. The first option is a permanently attached antenna. These antennas usually include devices requiring that the box be opened to remove the antenna. A nonstandard tamperproof screw secures the antenna to the box, the antenna is soldered to the box, or the antenna is molded into the radio.

The second option is that the antenna be professionally installed. However, the FCC's definition had been somewhat ambiguous. For the most part, high-gain antennas designed to be mounted on a building exterior or a mast generally fall under the professional-installation clause. It's generally understood that a "professional" is one who is properly trained and whose normal job function includes installing antennas. Several groups (Cisco, CWNE, and NARTE) offer certification programs for unlicensed wireless systems installers that would qualify an installer as a professional.

The third option allows a nonstandard or unique connector to secure the antenna to the transmitter. The standard clearly includes connectors such as TNC, BNC, F, N, SMA, and other readily available connectors. The usual convention is that the male connector has a pin in it and also has the threads on the inside. More esoteric connectors —such as MCX and MMCX or connectors that are similar to the standard connectors with reversed threads, nonstandard threads, nonstandard shells, or the gender reversed—are incorporated into Wi-Fi equipment. Some common examples are RP-TNC, RP-BNC, and RP-SMA. Basically, an RP-TNC chassis connector has a male core and a female outside. That is, the threads are on the outside of the connector, but the connector has a pin in it. The mating part has a female core and the threads on the inside. If one needs these non-standard connectors, rest assured, they are difficult to find. The best place to get them is over the Internet.

Lightning Protection, Grounding, and Bonding   It is important to properly ground any external antenna. Many volumes have been written about lightning protection, grounding, and bonding. Refer to these and the manufacturer's suggestions. However, if these are not provided, one of the key things to do is to provide an adequate ground through a ground rod. A ground lead should run from the rooftop antenna clear down to the ground rod with a minimum of bends in the line. The Ethernet connection should have a lightning arrestor on it that is connected to the ground system before going into the building. Also, it is helpful to put a loop in the Ethernet cable near the AP or bridge and near where it goes into the building.

RF Propagation   Relative to deploying Ethernet cable to install a wired network, RF propagation can be a difficult science. The following pages describe the engineering challenges related to installing a wireless network, particularly regarding limitations in range.

Multipath Interference   One of the major problems that plague radio networks is multipath fading. Waves are added by superposition. When multiple waves converge on a point, the total wave is simply the sum of any component waves.

Where two waves are almost exactly the opposite of each other, the net result is almost nothing. Unfortunately, this result is more common than one might expect in wireless networks. With omnidirectional antennas RF energy is radiated in every direction. Waves spread outward from the transmitting antenna in all directions and are reflected by surfaces in the area. Figure A-17 shows a highly

Figure A-17
Multiple wave paths in unobstructed field

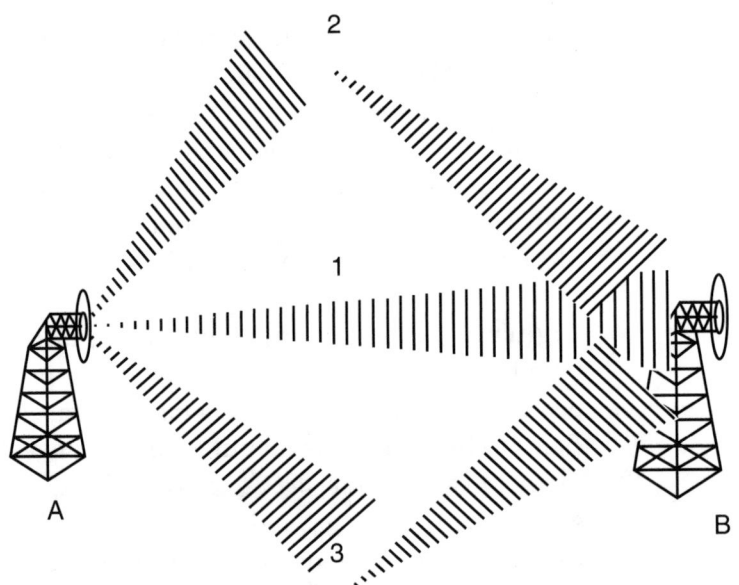

simplified example of two stations in a rectangular area with no obstructions.

Figure A-18 shows three paths from the transmitter to the receiver. The wave at the receiver is the sum of all the different components. It is certainly possible that the paths shown in this figure will all combine to give a net wave of 0, in which case the receiver will not understand the transmission because there is no transmission to be received.

Because the interference is a delayed copy of the same transmission on a different path, the phenomenon is called multipath fading or multipath interference. In many cases, multipath interference can be resolved by changing the orientation or position of the receiver.

Intersymbol Interference (ISI)   Multipath fading is a special case of ISI. Waves that take different paths from the transmitter to the receiver will travel different distances and will be delayed with respect to each other, as shown in Figure A-18. Once again, the two waves combine by superposition, but the effect is that the total waveform is garbled. In real-world situations, wavefronts from multiple paths may be added. The time between the arrival of the first wavefront and the last multipath echo is called the delay spread. Longer delay spreads require more conservative coding mechanisms. 802.11b networks can handle delay spreads of up to 500 ns, but performance is much better when the delay spread is lower. When the delay spread is large, many cards will reduce the transmission rate;

Figure A-18

ISI

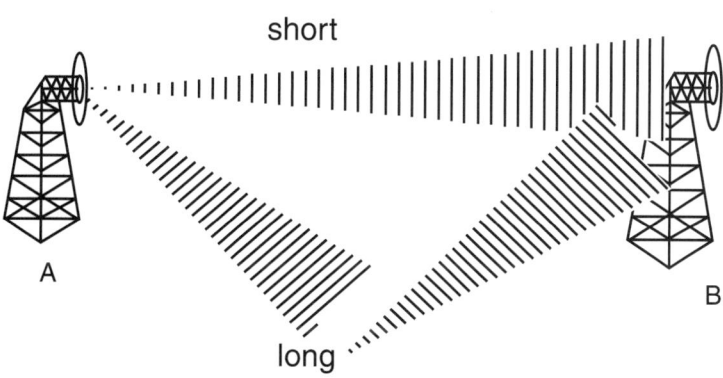

several vendors claim that a 65 ns delay spread is required for full-speed 11 Mbps performance at a reasonable frame error rate. A few wireless LAN analysis tools can directly measure delay spread.[6]

Using Two Antennas for Diversity   Diversity is often used with cellular BSs and is seen to help overcome multipath problems. Some BSs have two antenna connectors for diversity.

Anyone who listens to the car radio while driving in a downtown urban environment has experienced a momentary dropout or fading of the radio station at a stoplight. If the car moves forward or backward ever so slightly, the station comes back in. Although the car is in range of the radio tower, no signal is received in these dead spots. This phenomenon is called multipath fading and is the result of multiple signals from different paths canceling at the receiver antenna. Figure A-19 shows multipath cancellation from a large building.

Five different types of diversity can be used to increase signal reception in the presence of multipath fading: temporal, frequency, spatial, polarization, and angular. The first two types of diversity require changes in hardware.

Temporal diversity involves lining up and comparing multiple signals and choosing the one that best matches the expected time of arrival for a signal. This concept is implemented in some digital technologies. One of the most common methods to do this is adaptive equalization and RAKE receivers.

Figure A-19
Multipath
cancellation

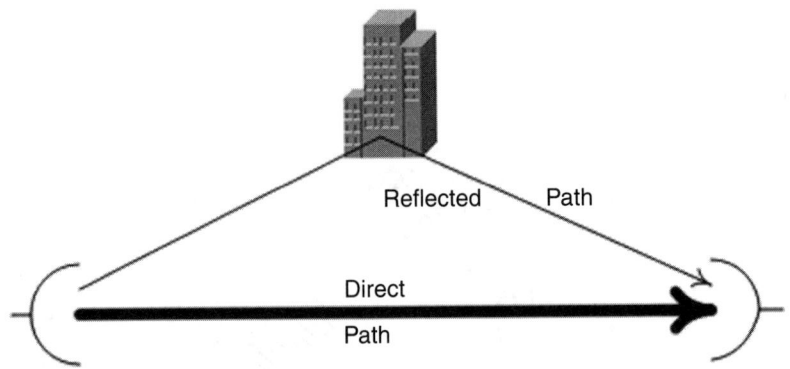

Reflected   Path

Direct
Path

[6]Ibid., 158–163.

Frequency diversity can be implemented by using two separate radio links on two different channels. If there is a null due to the cancellation of two signals because of a reflection, it will not happen on another frequency at the same place. Routers at both ends of the link could be used to send data across both wireless links. If one fails due to fading, the effective throughput is decreased. The redundancy of the link would also provide protection for other cases of failure.

Spatial diversity helps overcome the multipath problem by using two identical receive antennas separated by a fixed number of wavelengths. If there is a null due to a cancellation of the two signals, it will not happen at the other antenna. Because the antenna with the strongest signal is selected, the link is more likely to survive a fade when using spatial diversity.

The multipath problem can be helped with antennas mounted at *different angles* to cover the same coverage area. If a signal from one antenna and its reflection is cancelled, then a signal from a different antenna arriving at a slightly different angle will probably not cancel because the phase has changed. Again, because the antenna with the strongest signal is selected, the link is more likely to survive a fade when using angular diversity.

Finally, transmitting and receiving using two feed horns using both vertical and horizontal *polarization* (or clockwise and counterclockwise polarization) can also mitigate the multipath problem. When electromagnetic waves are reflected off of flat surfaces, their polarization can change. When the reflected wave and the direct wave combine to form a null, then had the wave been sent using the opposite polarization, no such cancellation would occur. Because the antenna with the strongest signal is selected, the link is more likely to survive a fade when using polarized diversity.

Weatherproofing It is important to seal all outdoor connections. But sealing has to be done in such a way that it can be removed if necessary. Use a combination of vinyl-backed mastic tape, heatshrink tape, and high quality electrical tape to seal the connector from moisture. Don't use silicon-based products or other spray-on or brush-on weather proofing materials. They are very difficult to remove.

How to Put a BS Where There Is No Power

In places where there is no power for the AP—such as inside a plenum or attic or on top of a roof—it costs about $800 to get an electrician to run power per code in addition to the $200 to run CAT-5 cable from the wiring closet to the AP. A number of commercial BS manufacturers have added Power over Ethernet (PoE) to their product designs to bring power over the spare pairs of the Ethernet cable to the AP (see Figure A-20). An injector is located in the wiring closet close to the power outlet. Their APs take power from spare pairs as part of their design. Also, a number of manufacturers are now offering PoE add-ons for most APs in the form of injectors and taps.

PoE also allows one to place the BSAP much closer to the antenna, thus reducing signal loss over antenna cabling. Ethernet signals are carried well over CAT-5 cable, but RF signals at 2.4 and 5.8 GHz are heavily attenuated over coax. Also, Ethernet cabling is much cheaper than coax. Figure A-21 illustrates PoE in a wireless network.

Figure A-20
PoE in an office

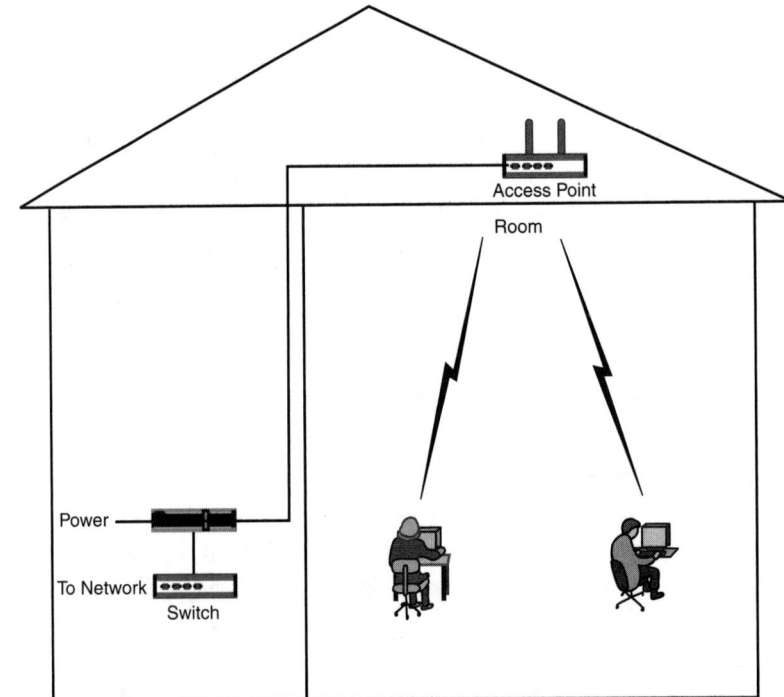

The wiring for PoE is relatively simple. See Figure A-22. Power is carried over pins 4 and 5 and pins 7 and 8. However, the polarity differs from one manufacturer to another. Most manufacturers use pins 4 and 5 to carry the positive lead and pins 7 and 8 to carry the negative lead of the power supply.

Besides the polarity, the voltages differ between manufacturer and by model. See Table A-8. The best practice is to stick with one standard and thus only the vendors of equipment that run on the same voltage and polarity. The IEEE is working on a new standard for PoE called IEEE 802.3af. The new standard will allow equipment from different manufacturers to sense the voltage and polarity of the power that is being supplied on the spare wires of the Ethernet cable and adapt to it.

Figure A-21
PoE in a wireless network

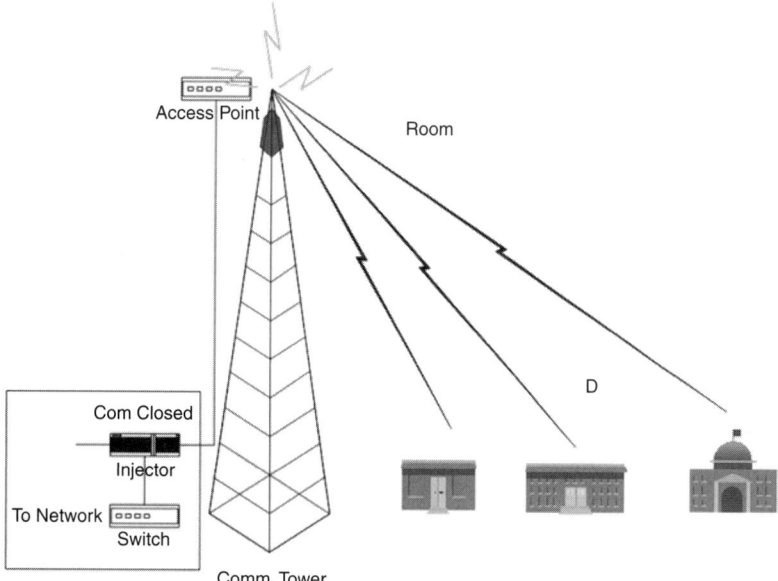

Figure A-22
Simplified schematic of PoE injector and tap

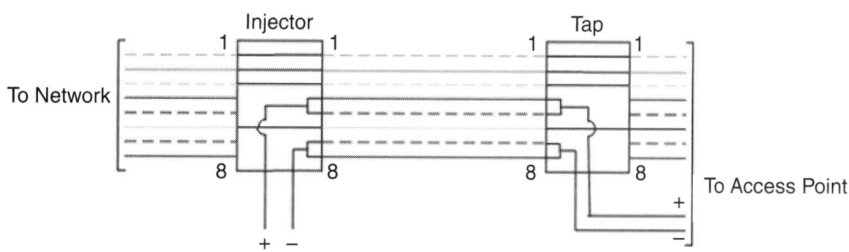

How to Overcome Line-of-Sight Limitations

The biggest challenge to providing Internet over WMAN is line of sight. So one of the keys to success as a Wireless ISP is to get sites with lots of height above average terrain or get hilltop locations and link those together using long-haul connections. From the key locations, it's possible to bring the signal to neighboring sites within a thousand feet or so. It's then possible to extend service from one location to a few more, as long as redundant paths are brought in to cover the new location. Perfect line of sight is not necessary when the signal is strong enough. Figure A-23 illustrates overcoming line-of-sight issues.

Table A-8

Injected Voltages and Polarities by Manufacturer

	Pins 4 and 5 + and Pins 7 and 8 −	Pins 4 and 5 − and Pins 7 and 8 +
5V		
12V		Intel, 3 Com, Symbol, Orinoco
24V		Intel, 3 Com, Symbol, Orinoco
48V		Cisco

Figure A-23
Tiered network to overcome line-of-sight limitation

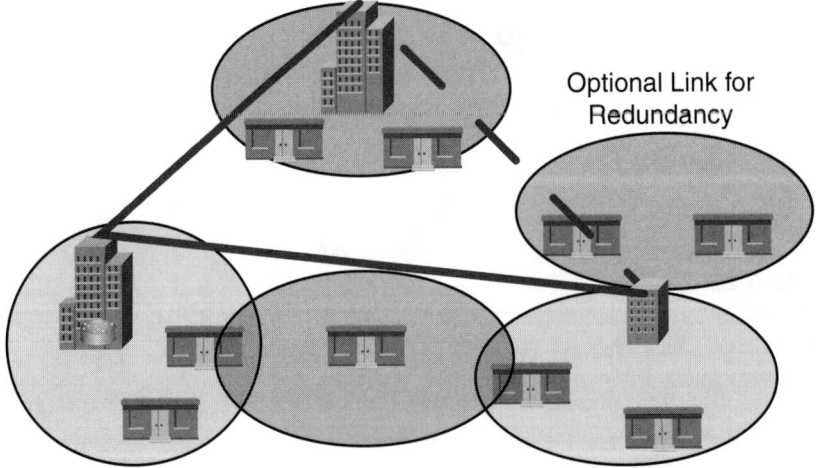

Optional Link for Redundancy

INDEX

S

X

Y